METAL STAMPING PLANT PRODUCTIVITY HANDBOOK

INDUSTRIAL PRESS INC.
200 MADISON AVENUE
NEW YORK, NEW YORK 10157

FEDERICO STRASSER

Metal Stamping Plant Productivity Handbook

Library of Congress Cataloging in Publication Data

Strasser, Federico.
Metal stamping plant productivity handbook.

Includes index
1. Metal stamping—Management. 2. Industrial productivity. I. Title.
TS253.S747 1983 671.3'3 82-15786
ISBN 0-8311-1147-X

First Printing

Metal Stamping Plant Productivity Handbook

PREFACE

Recently, productivity has become an important concept; everyone is now taking a greater interest in it. In 1978, the President's Council on Wage and Price Stability published a paper titled *The Productivity Collapse;* in 1979, President Carter in a speech to Congress cited declining productivity as a major contributor to inflation and strongly urged Congress to declare the increase in productivity of American workers a high national priority. It is quite correct that the appropriate people should be aware of and concerned with productivity, because through the improvement of overall production efficiency inflation can be combated, a high standard of living can be maintained, and jobs can be secured.

The change in the public's attitude toward productivity has been very encouraging. A few years ago productivity was considered a "dirty" word, denoting a way to exploit workers for the benefit of a few. Today, most people understand that increased productivity will benefit everyone, including the workers.

The title of this volume should not mislead the reader—the information given in many chapters is not restricted to metal stamping production; it is valid for any metalworking industrial plant and even for nonmetallic product manufacturing, for example, plastics. The field of production efficiency in the pressroom is enormous however; so within the framework and limitations of such a volume as this, it is impossible to deal with every detail with the same degree of thoroughness. Therefore, this volume will show the fundamental points the reader should consider and suggest directions the reader can follow to make additional improvements in productivity in individual plant settings.

The author wishes to express his appreciation for the valuable help and collaboration he received from many friends, colleagues, and manufacturers, and from the publishers, especially Mr. Ashak M. Rawji, Editorial Director, who helped greatly in the realization of this volume.

CONTENTS

METAL STAMPING PLANT PRODUCTIVITY HANDBOOK

CHAPTER ONE

Introduction

There seems no general consensus about the precise and correct definitions of the concepts of "productivity" and "production efficiency." A great many in industry speak about these concepts, and concientious factory managers aim to achieve what they think these concepts represent, but there is no general agreement about what productivity and production efficiency really mean. According to the majority of experts, productivity is a ratio of the *output*—in the form of production, i.e., goods or services (measured in quantities or in dollars and cents)—to the *input*—in the form of production factors and resources or, again, dollars and cents. In other words, it is the ratio of output in physical or financial terms to the factors involved in the output's production (i.e., labor, material, and equipment), and it represents the degree of efficiency attained. Consequently, optimum productivity signifies a high production efficiency, i.e., getting the largest possible and best quality output with the available and existing equipment, techniques, and space.

For some people productivity is intimately related to profits. This is based on the fact that by applying better production methods, the product unit costs decrease automatically. On the other hand, according to experts on basic cost-reduction principles, a dollar savings obtained through a productivity increase or improvement (i.e., through higher production efficiency) may be equivalent to as much as a ten dollar increase in sales. With the enormous cost of sales activities such an incentive should always be considered.

Speaking specifically about metal press production, the essence of productivity means manufacturing a high-quality item (component or finished article) with a minimum of handling, in the shortest time, and with the least input of labor, raw materials, energy, and capital.

The expressions "economical production" and "economy of production" are sometimes used as synonyms for productivity. However, these are not quite correct because it is possible to speak about "economical production" only as a comparison with other similar production methods, concluding that the one that offers a higher efficiency (a better utilization of the available resources) is more "economical" than the other.

It is the author's opinion that there are very few press shops in the world whose efficiency could not be improved. In every supposedly well-organized and well-managed shop producing metal stampings something can always be done to obtain manufacturing costs and manpower savings, to improve quality, to increase the production rate, and to speed up delivery dates. To achieve these goals, it is usually not necessary to abandon orthodox production methods and embark upon entirely new approaches, make revolutionary changes, invest large sums of money, employ a great deal of time and effort for studies, or prepare or provide special equipment. Of course, by acquiring new machinery for direct production or for material handling, more immediate and more spectacular results may be obtained. However, it is not always necessary to resort to such drastic means: often almost equally satisfactory results are obtained with simpler measures, which only require the application of sound practical principles, combined with ingenuity and common sense.

It is an interesting fact that all things start from small beginnings. And it is rather satisfying to note that in many cases through small, inexpensive methods or devices, amazing results may be obtained. Often great improvements have been realized by simply rationalizing the existing methods.

The installation of new sophisticated automatic production equipment can lead to higher production output; however, in such cases, shop management must face a new set of problems. In fact, with the extra investment, capital costs are higher, therefore, downtime becomes more expensive. In addition, sophisticated equipment often requires more-skilled labor, which is usually in short supply.

Improvements in production efficiency may be obtained in several ways: introducing simpler working methods (one of the basic goals of industrial engineering), improving the general physical working conditions, rearranging machines to get a more rational layout (straightening material flow, for example), modernizing existing equipment (installing new attachments, changing old machine members, rebuilding obsolete machines), redesigning tools, and, finally, acquiring new equipment, i.e., streamlining production-at-large by putting greatest emphasis on mechanization and labor-saving devices.

As anyone connected with the production of metal stampings knows, a great many details contribute to the success (or failure) of the operation of a press shop. Fortunately, the majority of these contributing factors are independent of each other, so it is not necessary to attack *all of them* simultaneously. Since productivity is a general term that covers several different production details, it is possible, and even recommended, to study each one separately in order to improve their individual contributions to overall efficiency. Even a modest program of improvements, dealing with a few factors only, is worthwhile: It will give tangible results and it will be possible to enlarge it as it progresses.

There are many separate items that should be studied thoroughly in order to see what can be done to increase production yield or to decrease rejects, taking into account the existing equipment and the possibility of acquiring new equipment. The approach to the problem with which we are dealing is quite different from that for short-run (low-volume) production. In order to avoid any misunderstanding, it must be stated that this volume deals with the regular mass production of stampings. Also, the approach will differ according to the size of the shop and the size of the workpieces in question. This volume deals with average conditions, i.e., with medium-size shops and with the production of medium-size stampings. Only cold working of sheet metal is dealt with; hot working is excluded, being somewhat of a special field. This volume is limited in its scope to the actual press shop. It does not cover stock preparation (strip shearing, strip slitting, etc.) or the ultimate processing or manipulation of stampings (finishing, plating, painting, heat treating, machining, assembling, etc.). However, the overwhelming majority of ideas and methods presented are universal, i.e., they may be profitably employed in any kind of press shop, only the means of their characteristics will be correspondingly different. Also, the improvements presented are applicable to both current production methods and new products.

To many people the word "productivity" evokes an image of a frantic workpace; actually, it is quite the reverse. It is interesting and gratifying to note that to increase productivity it is not necessary to put pressure on machine operators or other production people. On the contrary, with more efficient production methods press operators are liberated, partly or totally, from the slavery of frustrating, laborious, and monotonous repetitive operations; their job becomes limited to adjusting, checking, and controlling the automatic operations and the products. In this way unskilled workers are transformed into trained technicians with all the consequent benefits for them, their families, and the company they are working for.

When the suggested improvement methods and measures are adopted, the employees often work in a more relaxed manner and are happier with their jobs. It is the external, material, and mechanical circumstances that should be changed and improved. Rather simple methods are suggested in this volume that are often overlooked in everyday practice. As reported some time ago, a British company obtained an interesting and unexpected result after taking a party of schoolboys around the factory. During the subsequent talk around the tea table, some of the brighter children asked their guides why certain operations were done the way they were. Thus, fresh eyes had detected remnants of inefficiency that were so familiar to the staff that they were invisible. Such situations occur more frequently than one would think. Usually, the inefficient production conditions are established so gradually that they pass unnoticed, or some temporary, inefficient method inadvertently becomes permanent.

Only by constantly trying to improve performance, making full use of the existing manufacturing equipment, and applying the best techniques can better productivity be achieved.

The fight for better productivity is twofold: active and passive, i.e., to increase production efficiency (quantitywise and qualitywise) and to reduce spoilage (rejects and excessive scrap and waste). In fact, from the production efficiency standpoint, what matters is the total output in a given period of time of *usable* workpieces. Therefore, we are interested not only in keeping the output high, but also in trying to eliminate or keep as low as possible the number of unsatisfactory parts.

CHAPTER TWO

Plant Layout and Material Handling

As one of the very first steps in the program of production efficiency improvement, it is suggested that a close look be taken at the plant, with special emphasis on the pressroom, and at the arrangement of the equipment facilities, storage areas, and traffic zones. It will be seen immediately that the plant layout, itself, offers good possibilities for improvement, with little or practically no expense. In fact, efficient and economical operation depends to a large extent on the layout of the plant. Any company—small, medium, or large, with either easily or difficultly handled products—can profit from an adequately planned and executed plant layout.

Plant layout deals with the arrangement or disposition of the production factors—workers, materials, and equipment—within or near the plant buildings for the efficient processing and unhindered flow of materials in order to achieve highest production efficiency.

Advantages of Correct Plant Layout

Through proper application of correct and adequate layout principles, i.e., by judicious arrangement of the equipment and the production facilities, all plant spaces are utilized effectively and, consequently, important economical advantages are realized, such as, (1) increased overall production (output); (2) improved product quality, reduction of rejects and waste, better product and production inspection; (3) savings in time and labor costs (elimination of unnecessary overtime, decreased overhead); (4) savings in material (smaller material and part inventory requirements); (5) savings in material handling (through elimination of cross-handling and over-handling, which are wasteful); (6) savings in floor space or its better utilization; (7) reduction of associated costs by elimination of bottlenecks (through efficient use of equipment and elimination of crowded work areas and of inaccessible supplies); (8) better care of equipment (because it is easier); (9) easier employee supervision; (10) better housekeeping and thus improved employee morale; (11) improved safety conditions. In general, the actual capacity of the plant is increased, approaching its optimum potential.

This section deals, of course, not only with the rearrangement of existing plants but with the same principles and concepts which may be applied successfully to the planning of new plants.

Proper Method For Achieving Good Plant Layout

The means and procedures for reaching the goal of good press shop plant layout are rather simple. The fundamental principles are based on common sense and experience (like many of the suggestions contained in this book).

The first step consists of making a thorough study of the manufacturing methods and processes: Get process sheets or operational flow cards of the principal products; analyze operations and part flow. Then gather accurate detailed data about machines, storage, services, buildings, material handling, and other pertinent items. Next a rational layout is devised tentatively around the production process and the machinery needed for it, with first priority given to the three fundamental production factors: personnel (operators, die-setters, inspectors, etc.), machines (presses, attachments, shears, etc.), and material (stock, products, scrap, etc.). The next priority is nonproductive accessories—even if they are of an essential character—such as toolroom, maintenance shop, inspection, racks, cribs, storage areas, benches, washrooms, locker rooms, lunchrooms, first-aid rooms, offices, vending machines, which must, of course, be strategically and conveniently located and suitably integrated to the production facilities.

Since an efficient plant layout can be achieved only through teamwork, the first suggestion for a plant layout must then be discussed with everyone concerned with production: It is strongly recommended that the operators of the equipment that will be affected by rearrangement of the plant layout be asked for suggestions. Similarly, always consult (and try to get advice from) the person who will supervise the equipment and the people who will be involved occasionally—die-setters, maintenance people, safety personnel, etc.—since they have different requirements than the equipment operators. Finally, the definitive plan is prepared accordingly, taking in account every suggestion received.

Obviously, it is not possible to set hard and fast rules for good plant layouts—rules or standards that should be valid for every plant. However, there are some widely used details of general, practical, and proven value that should be taken in consideration in every plant layout study.

1. Referring to the building, preference should be given to areas (shops) without partitions and free from obstructions. Divided areas or separate rooms reduce space utilization efficiency.

2. Avoid congestions by organizing "traffic" to ensure proper "flow" of materials and products. Move materials and products between machines and/or storage areas as easily as possible and over minimum distances. Modernize material-handling methods and extend use of labor saving equipment.

3. Locate machines so that material can be deposited directly by trucks to be readily accessible at the side of the operator. Thus, there is no need for further unnecessary handling motions and movements by the operator.

4. Provide strategically located adequate storage areas for incoming and outgoing material (and for parts in-process). Suitable storage should be as near as possible to both supplies and utilizer machines.

5. Eliminate "hidden pockets" where material can be accumulated and

therefore, not accessible to material-handling trucks (pallet, fork lift truck, etc.).

6. Foresee the possibility for later expansion.

7. Make use of balconies for light storage and infrequent operations.

8. Put electrical conduit and air lines under the floor so that overhead handling equipment can operate unhindered.

9. Give higher priority to the steady flow of stampings and scrap than to the facilities of temporary and permanent storage for products. These latter may improve the appearance of the shop but may not decrease the production process time, which is what should be strived for constantly.

10. Check every detail for satisfactory working conditions (see Chapter 7) and then take them into account: lighting, ventilation, heating, safety, fire hazards, and air and power supplies.

11. The secret of a successful plant layout consists of *not* overlooking any detail. The layout of an ideal plant is presented in Fig. 2-1; it is laid out around the material flow and services.

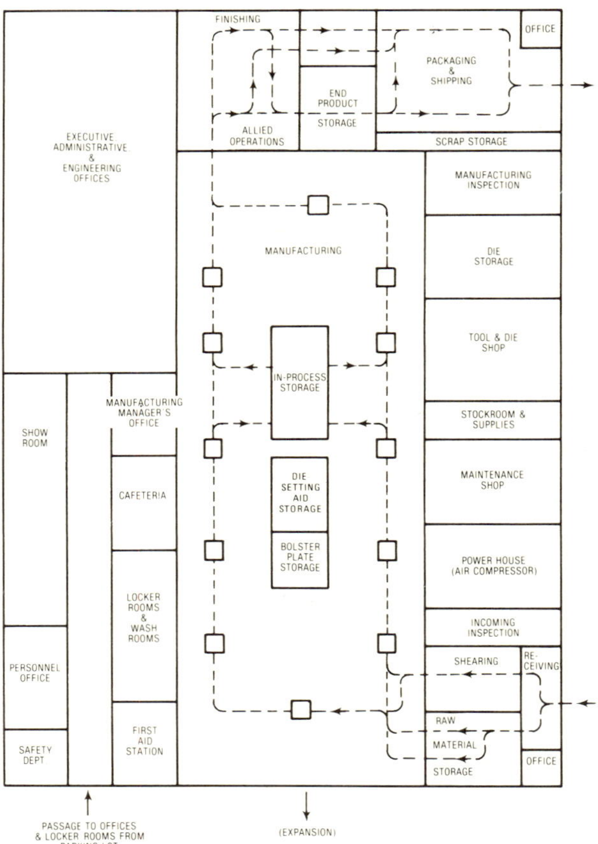

Fig. 2-1. Typical plant layout for mass production of metal stampings.

12. Finally, the definitive plant layout is always a compromise between "ideal" conditions and what can be achieved under the actual conditions of the plant.

Basic Plant Layout Patterns

There are two basic governing standpoints for internal arrangement of equipment (machines, racks, storage areas, etc.) according to the nature of the jobs usually performed in the shop: short-run, frequently changing production or long-run, steady-flow production.

In the first case the machines are grouped according to their own characteristics ("colony or type grouping"), such as type, size, function, stroke, etc. The chief advantages of this kind of arrangement may be summarized as follows: better utilization of equipment; lower capital investment, because fewer duplicate machines are needed for meeting production requirements; increased flexibility for changing the sequence of operation steps; steadier operation with fewer interruptions; and more possibilities for utilizing standby machines in case of emergencies.

The second case—process or product grouping—is used when there is a given manufacturing process to follow, i.e., a certain product or product group is to be produced in great quantities; then a functional layout may be planned. Such a rational arrangement is made by providing and following a natural, logical sequence of operations with the general unidirectional flow of material, through proper, reasonably designed "routes." The stock, enters at one side (or end) of the shop and "flows" (i.e., passes from one workstation to the next) always directly, following the shortest, easiest, fastest way (a simple, orderly, straight-line flow) until final inspection, shipping, or storage at the other end of the shop. Another variant of the same principle consists of entering the stock by a door, processing it around the shop, and letting the finished product exit by the same door. Every unnecessary detour, backtrack, crisscross, etc., must be strictly avoided not only because of increased handling costs and additional wastes of time and labor, but also in order to prevent the possibility of confusion, mistakes, misunderstandings, etc.

Further important advantages of the process grouping layout are as follows: (1) Reduced material handling—to incredibly low values—owing to the short, easy travel paths of the material; in some cases, material handling may be almost totally eliminated, when parts pass from one press to the next (located within a short distance) without any specific handling (using chutes, conveyors, etc.). (2) Reduced total production time. (3) Simplified and reduced production control costs and labor. (4) Since material in process is drastically reduced in quantity, less floor space, less capital investment, less confusion, and less congestion result.

For the case of product grouping layout, the geometrical scheme of material flow depends on the best (most economical, most efficient) operational sequence of the production process, on the size and shape of the building, on the characteristics of the machines and equipment involved, and on other secondary factors. Usually one of the following designs is adopted (or some combination among them): In case of very short processes, with very few operations, it is commonly a simple, straight line (rectilinear) (Fig. 2-2). For the case of longer processes (involving a higher number of machines and/or processing workstations) it may be U-shaped (Figs. 2-3), ringlike (annular) (Fig. 2-4), zigzagged (Figs. 2-5 and 2-6), comblike (Figs. 2-7 and 2-8), or combined (Fig. 2-9)

Fig. 2-2. Straight line arrangement of presses.

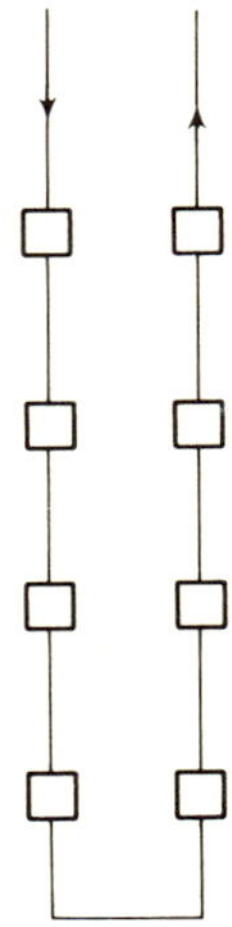

Fig. 2-3. U-shaped arrangement of presses.

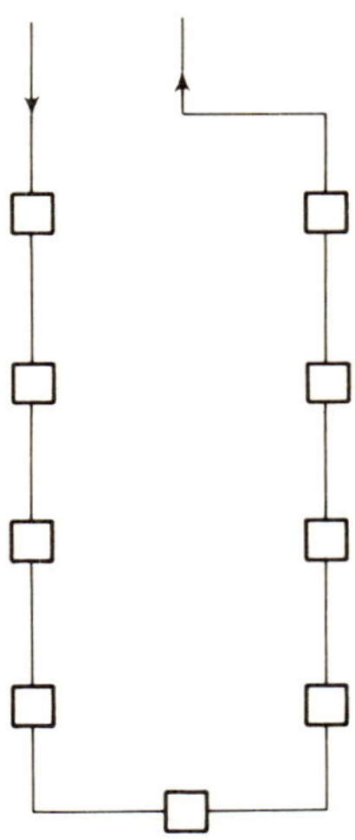

Fig. 2-4. Ringlike arrangement of presses.

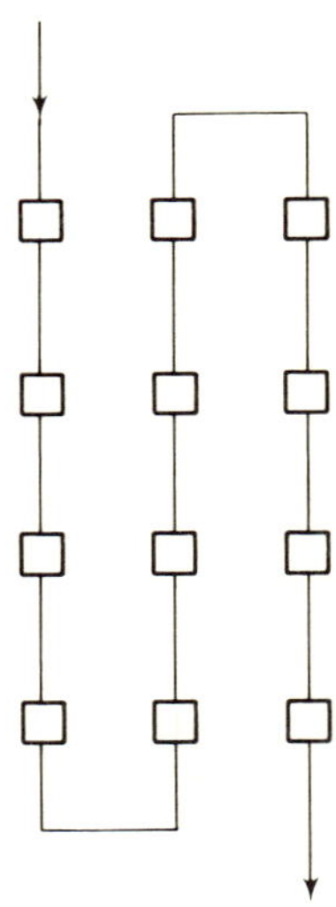

Fig. 2-5. Zigzagged arrangement of presses.

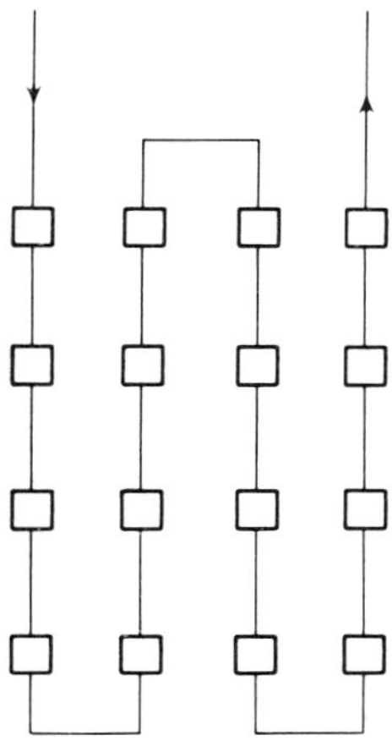

Fig. 2-6. Long zigzagged arrangement of presses.

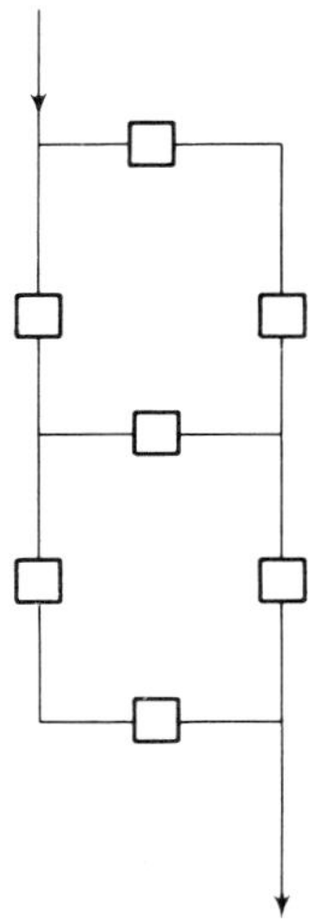

Fig. 2-7. Comblike arrangement of presses.

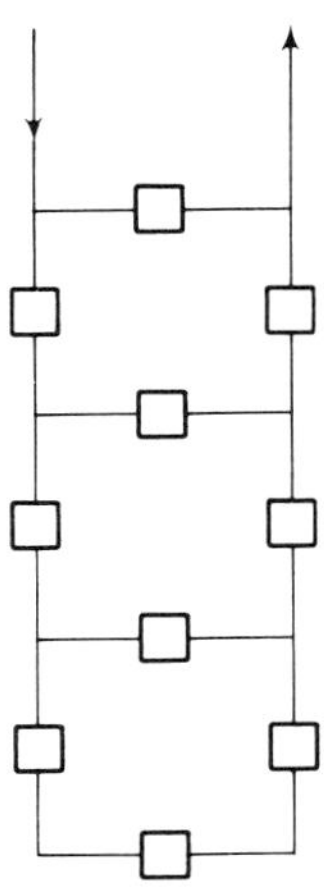

Fig. 2-8. Comblike arrangement of presses.

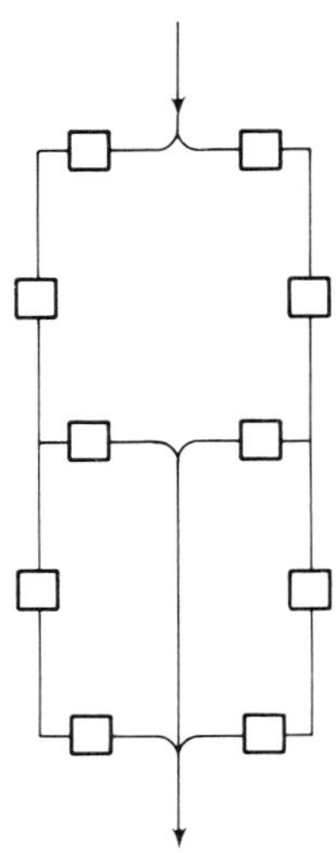

Fig. 2-9. Combined arrangement of presses.

Incidentally, the layout sample of Fig. 2-1 is a typical example of process grouping similar to the pattern of Fig. 2-7.

Aisles

Machines and equipment should not be allowed to occupy all the available floor space. Between machine groups, storage racks, etc., a well-planned net of "streets" (and, in some cases, even "plazas") called aisles must be provided and clearly delimited; these are usually marked by wide yellow lines. The aisles should be sufficiently wide so that stock- and die-handling trucks can be moved easily and freely to deliver material within easy reach of the operators. Such deliveries must be possible without disturbing nearby working operators.

The width of the aisles depends chiefly on the type and size of the material-handling trucks used in the shop. In the case of thoroughfare aisles the width must be such that two trucks can travel abreast with safety or a loaded truck can turn around. This means a width of 9–12 ft. In secondary side-aisles it is enough to ensure the passage of one truck at a time, without the need to turn around (loaded or unloaded), so a width of 5–6 ft is adequate.

Flexibility

Usually it is assumed that press shop layouts are more or less permanent; once a press is installed, it becomes an immovable object—it remains in place indefinitely. However, practical experience shows that in a surprisingly large number of cases there is a decided need to change the location of presses. There may be various reasons for this: expansion (market demands greater monthly deliveries); changes or modifications in product design, requiring either simpler or more-complicated processing; substitution of one product by another having more or less different characteristics; or introduction of improved or new manufacturing processes.

In order to avoid too much expense and delay, it is good practice to provide initially (preferably, during the first installation of a new plant) some suitable means for equipment relocation. These may be either continuous foundations for larger presses or boltless installation for smaller presses (see antivibration mounting of presses, below). In some cases very small, auxiliary presses are provided with caster rollers by means of which they may be moved easily.

Aesthetics

About the aesthetic aspect of the shop layout and equipment arrangement, opinions are very much divided. There are those who are not particularly concerned about plant appearance, only caring about function, efficiency, and economy. Others advocate a pleasant, good-looking, cheerful plant arrangement because of the indirect advantages obtained thereby. From the standpoint of employee morale and pride and for an attractive appearance when visited, by customers, suppliers, officials, etc., an aesthetically agreeable plant has some virtues. Incidentally, the author belongs to the group that prefers a good-looking working place.

Press Arrangement Patterns

Each press must be located so that the work can be done as smoothly and trouble-free as possible, and there must be provision for easy and quick press and die inspection, maintenance, and die setting, without any probability of disturbing nearby machines and operators. Press operators need agreeable and pleasant working conditions, especially adequate light and, most of all, ample space when standing or sitting, which means that sufficient space for stools must be provided. The operator's working position must be such that any person who passes or any material that is carried nearby, should not disturb him; in addition, anyone approaching him, should be readily visible, so as not to startle him by a sudden appearance (a safety consideration). At both sides of the press enough space must be left free for feeding and unloading devices, bins for blanks, stampings, scrap, tables for strips, etc. The places for incoming and outgoing material (stock, products, scrap) must be clearly determined and possibly also marked, and racks must be provided for accessories and tooling that belong (temporarily or permanently) to the press, in order to store them properly when they are not in use.

Press accessibility must be considered in every detail. The press disposition must be made in such way that the operator can reach his working place without having to make awkward maneuvers, which may cause accidents. In addition, the operator must be able to leave his working place quickly and safely in case of emergencies (accident to himself or to some neighbor, fire, etc.).

The press should be installed in such a location that it would be possible to move it without having to move other nearby presses, or at the most, only having to move one or two of them.

It is good practice to place presses under overhead cranes, so that heavy objects can be delivered or removed easily by the crane, e.g., supplies of coil stock or heavy dies. Also, in case of press repairs, revisions, or overhauling, the flywheel, bolster plates, and other heavy parts may then be removed easily and safely.

Floor space is expensive, both in first cost and operational expenses (building, depreciation, light, heating, maintenance, power for transportation). However, ample space often can be achieved around presses by proper orientation of the machines. In fact, by offsetting, i.e., inclining the press axis with respect to building walls or aisle center lines, the floor space is utilized more efficiently (analogously to screw machine arrangements). There are no fast and hard rules about the inclination angle between press axis and aisle axis; it may vary between 45° and 75°, with 60° being a popular average. With offset presses the stock can be handled more easily thus avoiding the danger that the strips, coils, or panels being worked in one press, would disturb the operator of an adjacent press. Besides, the removal of stampings and of scrap is made easier. A typical inclined press arrangement is shown in Fig.

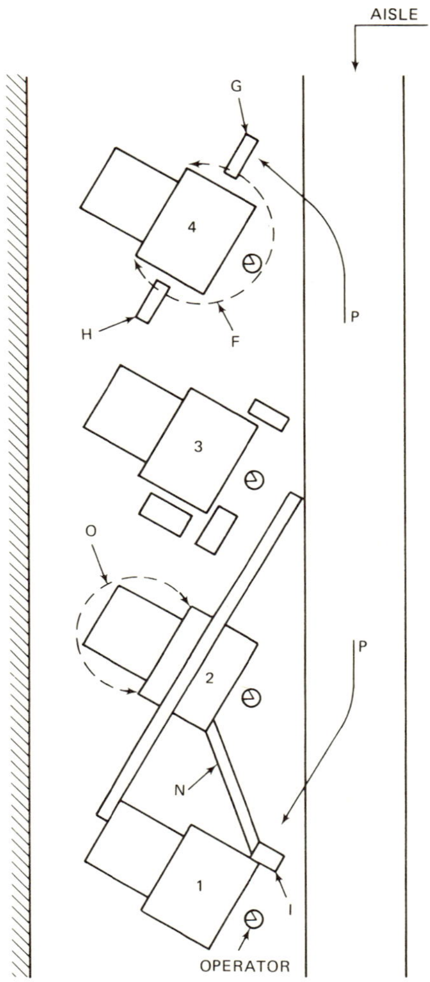

Fig. 2-10. Inclined press arrangement facilitates operations.

2-10. Ample freedom of action for the operator is illustrated by circle *F*. Space for press inspection and maintenance is shown with circle *O*. Arrows *P* show how the supplying of stock does not disturb operators of nearby presses. Each press performs different work: Press No. 1 has a second operation die, which receives from No. 2 the blanks through a chute (*N*) which fills a container (*I*). Press No. 2 works with strips. Press No. 3 works with blanks received from an outside source. Press No. 4 works with coils managed by an automatic feeder (*G*) and a strip skeleton winder (*H*). Each of the four presses does not interfere with any of the other presses.

It should be noted that such valuable advantages may be realized only when the

Table 2-1. Total Floor Space Needed for Presses Including Aisles and Other Space Required by Material Handling

Press Capacity Tons	Total Area (ft^2)	
	One-Point Presses	Two- and Four-Point Presses
5–10	30–50	—
15–50	50–70	—
60–100	80–100	140–170
120–200	110–150	190–250
220–450	170–200	280–350
500–1,000	230–280	370–450
1,200–2,500	—	470–560

presses are motorized, i.e., equipped with individual electrical drives. Obviously, antiquated overhead lineshaft drives have no place in modern plants.

The actual floor space needed for a given press depends on many factors, such as press size, shop size, press orientation, stamping size, lot size of stampings per run, type of operation(s) performed, and accessories provided for the press. As an average, 150 ft^2 may be taken as adequate for small and medium-size presses. Table 2-1 gives more detailed data in this respect.

Antivibration Mounting of Presses

Heavy and fast (high-speed) presses, when operating, produce noise, shock, and vibration that are transmitted to the floor and the building. This may be detrimental to the press operators, the presses, the tools, nearby machinery, and even the products. In order to control this problem, it is recommended that such presses not be anchored directly to the floor but indirectly by means of shock absorbers, which dampen the impact produced by the operating machine. There is a great variety of such antivibration mounting devices—cushioning pads, vibration mounts, vibration isolators. These devices employ some vibration-absorbing resilient material, such as cork, treated felt, natural or synthetic rubber (neoprene), steel springs, wood, or layers of fabric, vinyl plastic, or paper.

A satisfactory machine mount should be resistant to oil, grease, cleaning compounds, fuels, sunlight, dirt, organic acids, alkalis, and any other chemical agents that may be present on the floor of a metal stamping shop. Practically all such devices are commercially available products. There are several reliable manufacturers of vibration isolators who can give excellent advice about their products' use, and such devices are usually relatively inexpensive.

Figure 2-11 presents a selection of standard Micro/Level Isolators manufactured by Vibro/Dynamic Corp. Figure 2-12 shows Micro/Level Isolators being leveled and fine tuned. In Fig. 2-13 a press is being leveled and fine tuned with the Vibro/Dynamic LOD/SEN System.

Using antivibration mounts, several notable advantages are achieved, which—

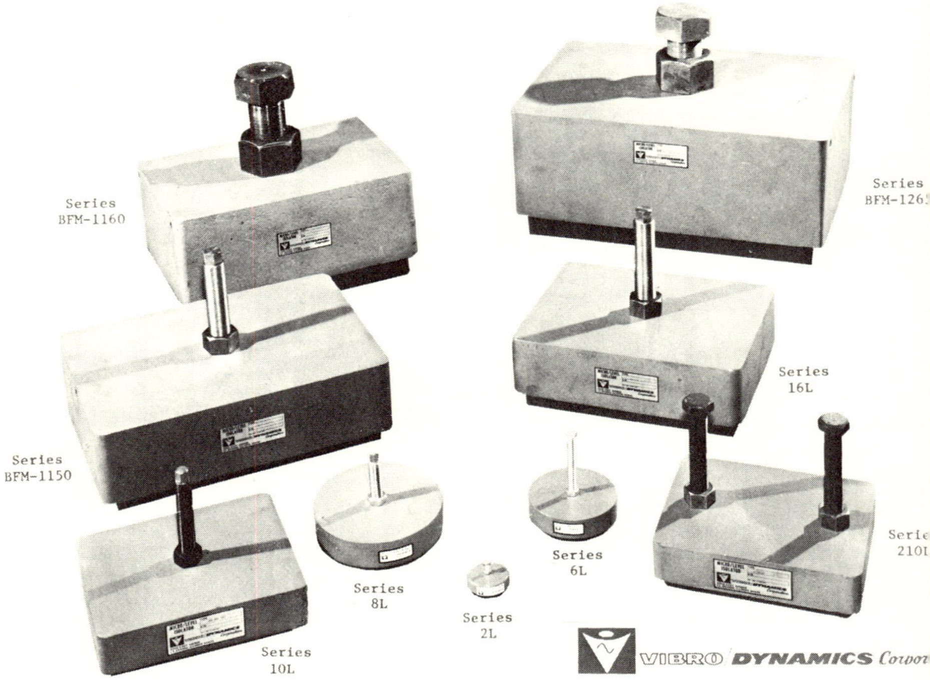

Fig. 2-11. A selection of Micro/Level Isolators. (*Courtesy of Vibro/Dynamic Corp.*)

directly or indirectly—contribute to the improvement of the overall efficiency of the production performance: quieter operation (less vibration and less noise); less operator fatigue; improved employee morale; improved operator efficiency because of the possibility of higher press spm; increase of production rate and output; reduction of power consumption resulting in a longer useful life of machines and of tools; fewer production interruptions; decreased downtime; reduced repair costs; improved product quality (less rejects) and precision; reduced installation and maintenance costs and time (not only of presses and tools but also of building and equipment at large); more efficient material handling (no need to locate the machines far from each other); possibility of plant layout for straight-line method—often the press is simply cemented to floor (no need for lag screws and expansion bolts); and more flexibility for rearrangement—they can be installed at any time.

Sometimes trouble is experienced even after installation of antivibration mounts. This is owing to the fact that piping (for compressed air and/or water), electrical connections (ducts), and other rigid accessories connected with both the machine and the building, continue to transmit vibrations. Such trouble may and must be eliminated by providing suitable flexible connections.

Do not use antivibration devices indiscriminately. Inclined and nonsymmetrically loaded machines may present especially delicate problems; therefore, whenever there are doubts, one should always consult a reliable manufacturer before installing these devices.

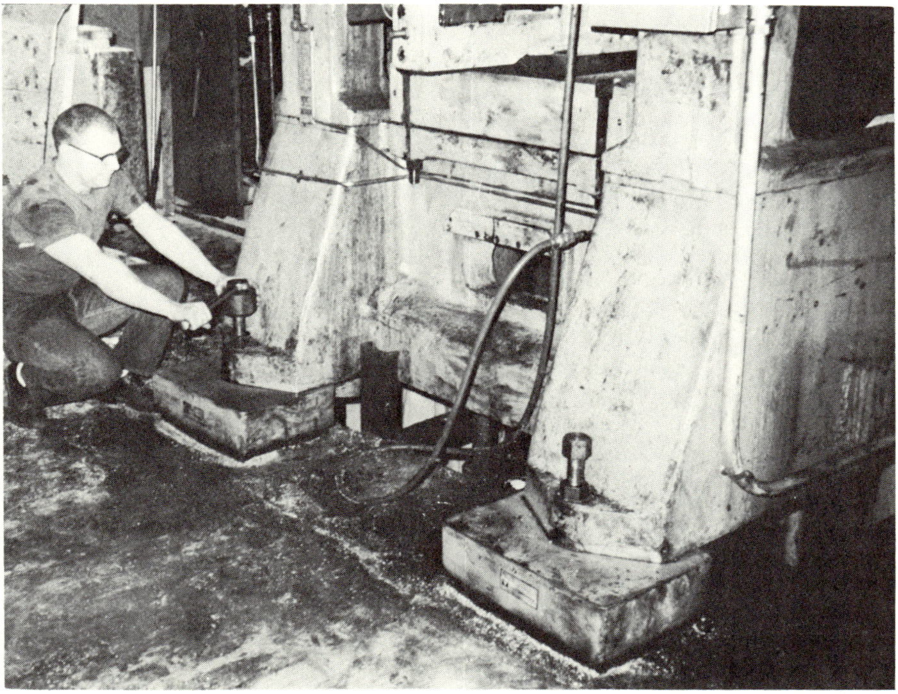

Fig. 2-12. Installation of Micro/Level Isolators and press leveling and fine tuning. (*Courtesy of Vibro/Dynamic Corp.*)

Building

The building in which the press shop is housed must be planned around the plant layout, i.e. around the chief manufacturing processes.

The overall trend today is toward one-story buildings, which have several advantages. First, it is easier to provide for the floor load. Since the presses and the adjacent die storage must develop rather high floor stresses (concentrated weights), it is good practice to install the machines on the ground floor. The second floor, if it is available, should be reserved for light machinery and for inspection, assembly, and similar activities.

The other great advantage is the fact that material handling is much less expensive when performed horizontally rather than vertically.

Other advantages of one-story buildings are: better space utilization, with no elevators or stairwells; greater freedom and flexibility for plant layout and rearrangement of machinery; less cost per square foot; quicker to build; more suitable for natural lighting; easier and more efficient supervision; and ventilation is straight through the roof.

On the other hand, multistory buildings have their own advantages such as: a greater economy in heating; the possibility of overhead storage; the use of gravity for material handling; and need for less land for the same amount of floor space.

Fig. 2-13. Press leveling and fine tuning with LOD/SEN System. (*Courtesy of Vibro/Dynamic Corp.*)

In any case, when determining the size of a new plant, always keep in mind the possibilities of future expansion. There is no greater enemy of production efficiency than cramped working space.

Material Handling in the Press Shop

For best production efficiency, the machines in the press shop should have the highest time utilization factors possible. This means that, ideally, they should never be idle between die changes and maintenance operations. One of the principal means for attaining this goal is efficient material handling organization.

Modern material handling is an integrated production tool and as such has an important influence upon production efficiency as a whole, by helping proper material flow, preventing accidents, and reducing damage to products. The main contribution of material handling to productivity is indirect, by keeping the handling costs as low as possible.

Material handling consists of two distinct activities or functions: dynamic and static. The dynamic part means material flow, that is (a) transportation with manual means; (b) transportation with mechanical means; and (c) manual manipulation. The static part of material handling refers obviously to storage.

Even a superficial observation clearly shows that in the press shop material is handled for a much longer time than it is actually processed. The raw material has a considerable way to go until it is transformed into a formed or stamped workpiece.

Much handling, manipulation, lifting, and transporting must be done during this process.

It must be stated emphatically that material handling consists of not only the transportation of great masses of stock at large, but also any displacement effected by operators. And it is interesting to note that just this latter—which is sometimes only a travel of a few inches or even a fraction of an inch and which could be called "microhandling"—is at least as important as the "macrohandling," i.e., handling of great masses for long distances. In fact, this macrohandling means moving greater masses, but at long intervals, while microhandling means short movements, but multiplied by a large number of times. So, the product weight × quantity × frequency is usually much higher for the case of microhandling than for the case of macrohandling.

In addition, there is also another circumstance to be taken into consideration. The incoming material (bundles of sheets, coils, etc.) and heavy objects that require adequate handling equipment. The stamping process divides this incoming material into short pieces that are transformed into stampings and scrap. Consequently, these require different types of material handling equipment, adequate for light, small, but comparatively voluminous objects.

Adequate and economical material handling is achieved by two basic conditions: correct selection of efficient material handling means and their effective utilization, which implies planned material handling operations.

Material handling in the press shop may be divided into a few groups of rather different characteristics, goals, and means according to the nature of the activity: constant, frequent, occasional, and exceptional material handling.

The stock of raw material or preprocessed stampings must be fed *constantly* (with regular intervals of a few seconds or fractions of a second) to the dies for the stamping operations.

Stock must be supplied to the presses at determined and regular intervals, and finished stampings and scrap must be removed *frequently*.

Tools and dies are only *occasionally* handled, as when they are changed or there is some breakdown or maintenance job to be performed.

Finally, in some *exceptional* cases, when process changes must be effected, the presses are moved to rearrange their location.

The approach to the four groups is basically different: In the first two groups the chief goal is to save time and labor, while in the last two, the chief goal is to avoid hazards to the die setters, toolmakers, floormen, as well as to the equipment.

It should be noted that, basically, material handling comprehends every transporting and every moving of raw material, product, or scrap that happens in a plant. However, in the metal stamping shop, according to an arbitrary custom, it is understood that to material handling proper belong only those activities which are performed by material handling personnel, called floormen, while material transportation or movement (i.e., "microhandling") that is performed by press operators is considered as feeding or unloading. According to this unwritten convention, this section deals with material handling proper, i.e., stock, product, and press handling, while feeding and unloading of press tools are covered in Chapters 10 and 11.

In the press shop a minimum of physical effort should be required for the handling of stock (sheets, strips, coils, blanks, stampings, and scrap) and equipment in general. Mechanized equipment (powered hand trucks, self-propelled power forklifts, pallet trucks, conveyors, etc.) should be used most extensively in order to reduce manual handling to a minimum (especially if it requires heavy physical effort). Manual handling of raw material and/or material in process is most expensive because it is labor intensive.

The actual stock handling costs in the press shop represent a surprisingly high percentage of the average cost of a stamping. Such handling costs usually pass unnoticed; they are not reported separately in cost control accounting. They are dealt with simply under the heading of "burden" or "overhead." In most cases people do not realize the basic axiom that "Handling adds nothing to the *value* of the product, but increases the *cost*." It should be kept in mind that handling time and cost reductions, contribute directly and immediately to an increase in production efficiency and to the improvement of production economy.

The general lack of accounting about the magnitude of material handling costs explains this paradoxical situation: There is usually an enormous disproportion between the total material handling labor costs and the total investments in material handling equipment. In fact, it is imperative to provide every press shop with abundant mechanical transportation and other material handling pieces of equipment so that the labor costs for material handling can be kept as low as possible.

Several interesting and important advantages are obtained through adequate, up-to-date, and improved material handling methods.

The chief *goal* of efficient material handling in the press shop consists of trying to reduce the time spent by the material in the shop. Other objectives and advantages of correctly planned material handlings are:

Reduced volume of material in process

Reduced indirect labor costs (reduced manpower overhead)

Increased capacity of the existing building (less of a need for communication roads, paths, gangways, etc.) enabling space to be used for more production work; Also made possible by effective use of "third dimension", i.e. space above floor level.

Improved utilization of existing production equipment (creating faster working by speeding up flow of material between operations and processes)

Increased volume of output

Improved inventory control (no time lost looking for mislaid stock)

Enables operators to work at a consistently high level of productivity because it reduces unnecessary handling fatigue and allows efforts to be more productive

Better overall housekeeping

Promotes greater industrial safety (reduced accident rate)

Improved morale as a result of their operating mechanized equipment instead of doing hard manual work

Detailed study of material handling encourages and in many cases enforces improvements of plant and building layout

Improved product quality

Less damage to finished work in storage and work in process

Less turnover of skilled productive labor

Reduced cost of transportation

Increased capacity of distributive transportation

Reduced loading time and cost at the shipping point

Reduced overall time or labor, especially at the receiving point

Enables shorter delivery times to be achieved

Enables more flexible schedule of production

Plant is more attractive to visitors

The material handling equipment used should permit production operators to spend all of their time in actual production, handling material only at the press. (Bring the work to the workers; avoid the need for them to go after the stock or parts to be processed. Everyone should be aware of the fact that idle machines cost a lot of money.) Such an efficient organization of material handling demands certain conditions to be fulfilled: The stock to be stamped must be: (a) prepared on schedule; (b) delivered in the right quantities to the operator; (c) delivered to the right place; and (d) delivered on the first move.

"On schedule" means before it is actually needed. Material handling must be carefully planned beforehand as an organic and integral part of the overall production plan. It makes no sense to refine the operational time of presses and to find methods and means for reducing the active labor time per operation, if the press operator has to wait for the material handler to remove scrap and parts or to bring him fresh raw material to be processed. The press operator should never wait for stock, it is more economical for the handling equipment (and people) to wait for the operator. Supplies can be replenished between shifts or during coffee breaks or at such regular intervals that cause least interference with production.

"Right quantities" is a very elastic concept. Every case must be considered individually and decided according to its characteristics. In the case of small stampings and short run jobs, the stock necessary for the whole run (or at least for a long period of time) may be considered as the "right quantity." On the other hand, large stampings take up so much space that the supply of stock to be delivered each time to the operator should be only a small, quantity (in order to avoid congestions of material, thereby creating safety hazards and lowering production efficiency). On the average, it is recommended that the operator be provided each time with a supply of stock for at least two hours of work.

"Right place" means that the material must be delivered within easy reach of the press operator. It should not be necessary for him to leave his operating place and go behind the press or to the side of the machine, or elsewhere to get new stock or to remove finished work. If it is unavoidable to remove the finished stampings at the rear of the press, then proper helpers should be provided. Also nothing should pile up around the press because that increases the possibility of damage, losses, accidents, etc.

"On the first move" means that the floorman should locate the stock in the place from which the operator picks it up (one by one) to introduce it into the die. The material must be kept as close to constant motion as possible. Congestion must be avoided. Only if absolutely necessary should suitable intermediary storage spaces be provided between operations for parts in process. Again, the greatest emphasis should be put on proper flow of stock, products, and waste. Well-kept storage spaces may look ordered and give the impression of good housekeeping; however, they do not improve handling and therefore overall efficiency. On the other hand, if a process has to be stopped for a long period of time, it is better to move the parts in process from the production area to proper storage rooms than to let them lay around shop where they interfere with its normal operation.

Material Handling Equipment

Owing to the wide variety of stampings, of shop sizes and plant layouts, and of primary and secondary working conditions, it is impossible to give even an approximate outline for the best equipment and procedure for stock and equipment handling that would be useful for a large number of shops. However, the fundamental tasks—receiving the sheets, storing them, slitting them in strips, passing the strips through the shop to produce the stampings, collecting the parts and the scrap, and shipping the former and disposing of the latter—are similar in press shops even if they each have quite different characteristics. It is interesting to note that recently the design and development of material handling equipment has been improved considerably so that for every type of shop there is available equipment for stock and equipment handling which would notably improve handling conditions. This section gives a general idea of which kind of equipment may be commonly used for improving the efficiency of each phase of material handling in the press shops, but material handling equipment representatives should, of course, be consulted.

Handling of Sheets

Metal sheets arrive from the mill or from the warehouse loose or bound together in bundles. They may be handled during their movement from the delivery truck to storage and from there to the squaring shear, where they are divided into strips, or directly to the machines (presses, press brakes, etc.) in several ways: manually without any handling equipment, with some muscle-powered mechanical aid, or with fully mechanized means.

No matter which transportation method is employed, three basic conditions must be satisfied during the handling of sheets, and, for that matter, for handling anything in the press shop: *safety* to the people and safety to the equipment, *care* of the stock (damaged, bent, or even scratched sheets mean spoilage or at least additional cost for removing these defects by grinding, polishing, etc.), and *economy* (the apparently cheapest method quite often turns out to be more expensive than the others).

With the old-fashioned, inefficient, totally manual handling method, it is absolutely indispensable to provide some protective device, in order to avoid hazards to the hands and fingers of the floormen transporting these sheets. Such an accident or injury preventing device may be a proper glove or a leather or rubber disk provided with a hole for the thumb (Fig. 2-14) or bound to the wrist (Fig. 2-15). In other instances a properly shaped hook (made from round steel rod) is a great help (Fig. 2-16). By providing the hook with a roller (Fig. 2-17), the transportation is greatly facilitated. (These obsolete methods—which should not exist in modern, progressive plants—are mentined here chiefly for historical interest.)

There is an additional important reason for using gloves, in general, while handling sheets. Namely, if the sheets are handled with bare hands, perspiration (which is often quite corrosive) may leave unseeming, rusted imprints and thus ruin the surface finish of the sheets.

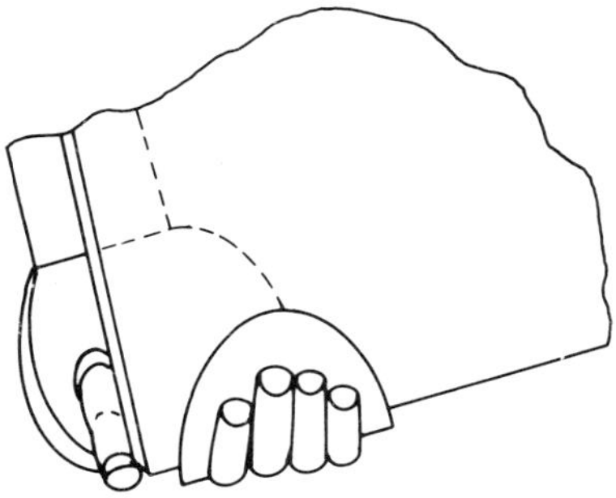

Fig. 2-14. Simple aid for manual handling of metal sheets.

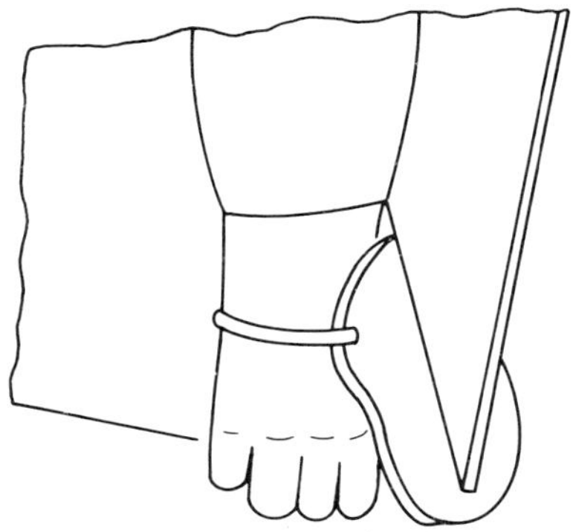

Fig. 2-15. Simple aid for manual handling of metal sheets.

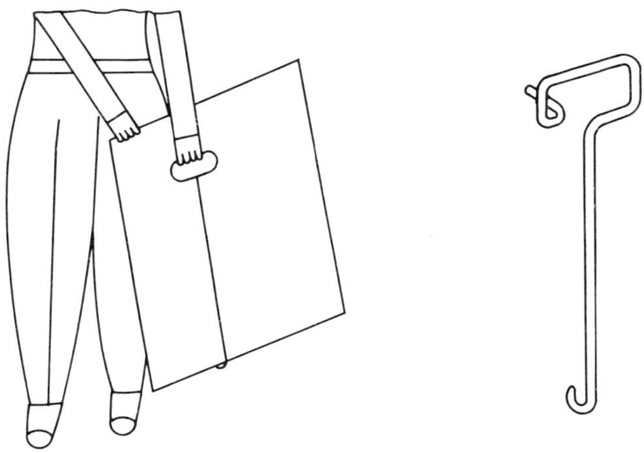

Fig. 2-16. Special hook for metal sheet transport.

Fig. 2-17. Sheet transportation hook provided with roller.

Manual handling is slow and inefficient; whenever possible, sheets should be transported with wheels, by mechanical means. Putting the sheets either bundled together or loose upon a hand lift truck or a mobile rack with casters is a great improvement, even if the truck or mobile rack is moved by manually. Incidentally, it is often easier to handle the sheets if they are arranged vertically (Fig. 2-18).

By making use of labor-saving equipment, i.e., replacing hand-driven trucks with power-driven trucks, putting the sheets upon heavy wooden skids or pallets and sometimes providing them with casters and thereby converting them into a type of trailer, the material handling efficiency is thus greatly improved.

It should be stated that power-driven trucks may be employed not only in large and medium-sized plants, but also in small-sized plants. There are interesting reports of case histories where a single power-driven truck has been used for all the transportation needs in a stamping plant (sheets from storage to shears, strips from shears to presses, product containers and scrap boxes from presses to destination—shipping or storage—dies between storage and presses, etc.). In all respects, such an application has been successful with consequent improvement of production efficiency, economical convenience, employee morale, etc.

Occasionally, in medium-size and large shops, sheets are transported with cranes, hoists, or lifters, after they are suspended with proper hooks indirectly, i.e., making use of skids or pallets. In other cases, the cranes and other overhead conveyors work directly, i.e., making use of magnetic or vacuum-operated suction cups. However, it seems that in the press shop processing sheet metal overhead transport is of less importance than floor transport, except in shops which produce comparatively large stampings (for automobiles or airplanes, for example).

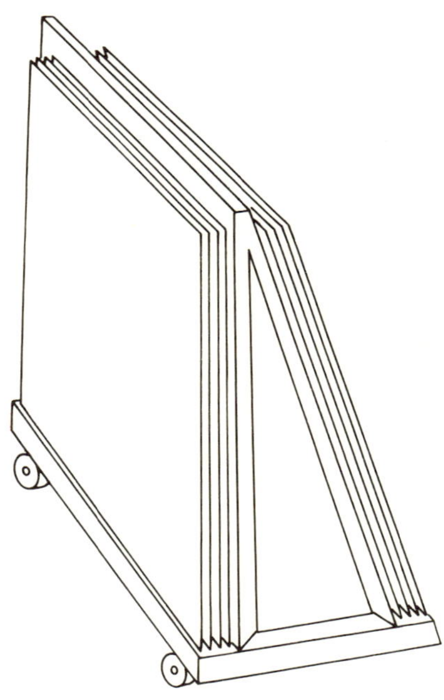

Fig. 2-18. Vertical arrangement of metal sheets, on a trailer.

Storing of Metal Sheets

Sheets should not be kept outdoors. If this is unavoidable, then they should be protected from the elements with waterproof cloth or impregnated paper or other suitable materials.

Sheet storage buildings or rooms must be clean and dry and have good ventilation; in winter, they must be well heated at constant temperature. This temperature should always be kept *higher* than the outdoor temperature. If the sheets are warmer than the incoming fresh air, then the humidity contained in the air will not condense on the sheets, and thus the danger of rusting is avoided. In addition, if the sheets must be kept in storage for a prolonged period of time, they should be properly coated with a lubricant, usually with some completely neutral light mineral oil which serves as a protection against oxidation.

Sheets and bundles of sheets should never, under any circumstances, be kept directly on the floor or on the ground. They must be so stacked that with the aid of proper wooden separators the forks of power-lift trucks can enter the spaces between the stacks (Fig. 2-19). It is also recommended that the sheet bundles be provided with proper protecting squares where the slings and chains hold them for transportation.

Soft metals and those destined for drawing and forming operations must be handled with extreme care; they must be kept perfectly clean and without scratches and not subjected to bumps, blows, etc. Careless handling of such sheets may have far reaching harmful effects. In fact, a sheet covered with dust and dirt, when pulled out from the stack, easily becomes scratched and it may scratch other sheets. Thus spoiled stampings, especially drawn shells, are produced that become rejects. Even the corresponding drawing and forming tools are easily damaged in this way; surface imperfections are created which must be eliminated causing production interruptions, extra tool redressing costs, with time and labor spent unnecessarily.

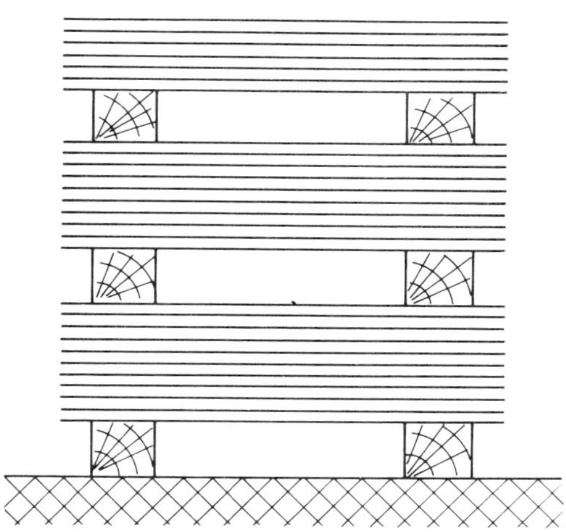

Fig. 2-19. Horizontal storage of metal sheets.

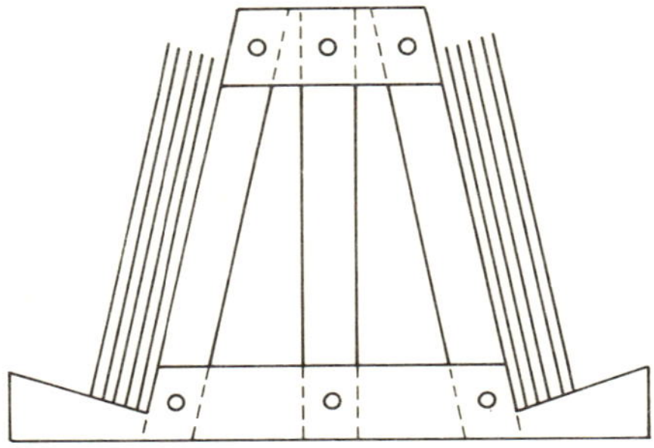

Fig. 2-20. Vertical storage of metal sheets.

The best way to store metal sheets is to hold them vertically (actually an inclination of 60–70° has been found most suitable) in proper racks, and not horizontally piled one upon the other. Figure 2-20 illustrates the principle of easy vertical storage which permits taking out any sheet in the middle without the necessity of rearranging the whole stack of sheets, as is the case with horizontal storage.

Although vertically storing the sheets requires more space (storage rooms cannot be easily exploited vertically), it should be preferred because of the advantages it offers. This means that it is easier to practice the principle of "first in, first out," which should be striven for in storing at large, thus avoiding such disagreeeable problems as the consequences of age hardening.

Sheet Feeding Tables

The transportation of stock in the press room as a typical material handling operation and the feeding of the dies are so intimately linked together that sometmes they overlap. As was mentined previously, feeding is essentially a materials handling operation. Thus, there is specially devised equipment which is used for both transportation and feeding (stock positioning) purposes: for example, metallic tables with casters which permit easy movement.

In order to always keep the stock at the correct working and feeding height, the table level is adjustable with telescoping or "scissors"-like movement. For this purpose, manual power sometimes is employed (usually actuated by a foot pedal, occasionally by a hand lever), but more often mechanical (hydraulic or electrical) power is employed as in Fig. 2-21 (See also Figs. 10-9 and 10-10). Thus operator fatigue due to lifting is almost completely eliminated; consequently, worker morale is improved and efficiency is increased. (The same portable elevating tables may also be used for the handling of dies.) In the few makes of sheet handling tables, the table top level position is regulated automatically so that the top sheet is always at the same preestablished level, no matter how many sheets are below.

In some cases the sheet feeding table is stationary, i.e., *not* supplied with the wheels or casters. Thus it is used only as a feeding attachment; in such cases it is transported by trucks. It is usual to provide several shelves so that a quantity of stock may be stored under the table top. Sometimes, as when the table is used for handling strips, there may be an oiler attachment for the automatic or semiautomatic lubrication of the stock. There are also sheet feeding tables which allow the table top to be tilted in order to make the feeding operation easier in the case of inclined presses.

Fig. 2-21. Sheet feeding table. (*Courtesy of American Manufacturing Company, Inc.*)

Strip Handling

Sheet feeding tables, both movable and stationary, are also used for the transportation and feeding of sheared strips. These tables are sometimes narrower than feeding tables for sheets.

Coil Handling

In the majority of cases of permanent tooling mass production, the stock is in the form of coils, instead of sheets which must be sheared into strips. For coil handling practically the same kind of equipment is used as for sheets with the exception, of course, of sheet feeding tables.

Usually fork lift trucks with or without some special attachments are used for coil transportation. If the presses are located under an overhead crane or hoist, it may be used for coil handling. For this purpose, the usual auxiliary means (slings, C-hooks, coil grabs, tongs, chains) are used. In addition, as press attachments, coil cradles and reels are employed from which the coil is unwound in order to feed the stock into the dies.

Handling of Stampings

The end products of press shops are stampings and scrap. In a well-kept press shop nothing is left on the floor, because picking up stampings, punchings, and scrap means unnecessary work, which costs money and time.

There are several ways of handling stampings, according to size, shape, surface finish, quantities, etc. Apart from the characteristics and specifications of the parts, the handling itself may be effected manually, automatically, or mechanically. Also, the ejection method (manual, semiautomatic, fully automatic) has a decided influence upon the further handling of stampings (see Chapter 11).

In the case of small stampings, the most usual method consists of collecting them in proper containers, tote boxes, bins, etc., directly as they exit from under the dies. When the size, shape, and other characteristics of the stamping, in relation to press table arrangement, do not allow collecting the stampings directly below the press bed, then the use of gravity chutes (with or without top covers), which conduct the stampings laterally or to the rear, is recommended. In the case of multiplex dies it is recommended that separate chutes be provided for each blanking station. Thus, the periodic inspection of the stampings and tool is made easier and reveals immediately which station is the cause of any trouble that might arise.

Vibratory chutes are recommended for those cases where gravity alone is not sufficient to move the stampings properly. Such cases are characterized by (1) low slope; (2) light gage stampings, especially if small (low weight); and (3) heavy lubricant used for the operation.

In other cases special conveyors are provided which may be driven, intermittently or continuously, directly by the press (rarely) or which may be self-propelled (usually). These conveyors may not only remove the stampings and scrap from under the press table, but, if necessary, may also elevate the parts, depositing them at a greater height than the press table itself (for example, interoperational handling; see Fig. 11-18). In any case this avoids the necessity for the press operator to collect the stampings periodically and to transport them to the tote boxes.

It is very convenient for the trouble-free and efficient handling (transportation, storage, and further processing) of products and scrap to use standardized bins and tote boxes. One may use either a manufacturer's or the plant's own standards for such containers. Obviously, the size and shape of the bins are determined by the size of the stampings and the quantities of the lots to be held in each container. Abundant use is made of barrel-shaped drums, but in the majority of cases the rectangular

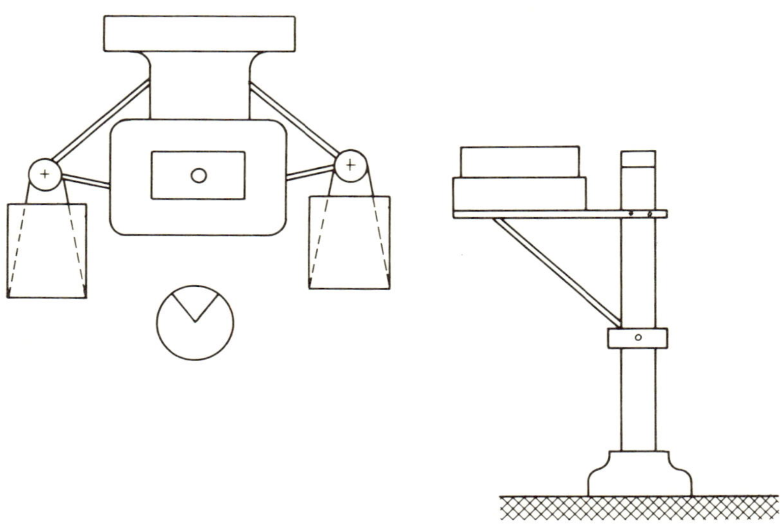

Fig. 2-22. Small containers for stampings and scrap within easy reaching distance of the operator.

shape is preferred because better stacking and a higher space utilization factor are possible.

Boxlike containers should have their dimensions determined with reference to the respective storage rack or shelves (or vice versa, the storage racks or shelves should be dimensioned according to the standard bins selected). In width, height, and depth the bins should fit tightly into the available space provided.

Large bins are usually provided with short legs so that they may be put directly on the floor, and the forks of trucks may reach below to lift them. Middle-sized and small containers may be put on pallets or skids or they may be put on spindles (bolted on the floor or even at the sides of the press) with swinging arms which support the containers within easy reach of the operator (Fig. 2-22). The correct position is attained by means of the swiveling arrangement of the arms. Finally, for small workpieces the bins may be mounted on the press, itself.

Drop bottom boxes considerably reduce the handling of stampings. Such boxes, filled with preprocessed blanks from a previous operation, are placed in an easily reachable location; the hinged portion of the drop bottom is then opened, thus allowing the workpieces to slide forward, making them easily accessible to the operator.

It is good practice to design small and medium-sized bins and tote boxes so that they can be stacked easily. Thus, vertical space may be utilized to full advantage. In other designs the empty boxes can be nested together (one entering the other; tapered walls allowing such arrangement) thereby occupying very little space.

In order to get the best productivity, every bit of material handling that can be eliminated should be. One such instance is the pouring of stampings from one box to another. Containers must be designed so that the bin which receives the stampings or small scrap directly (or indirectly) from the press should also be used for any further handling purposes, such as transportation, storage, and assembly; in some exceptional cases these containers are even used for shipping.

Handling of Slugs

In the case of progressive-type dies—where in some preliminary stations small slugs are punched out—while the stock passes through, the stampings and the punching must be collected together and separated with sieves afterward in a different operation. These sieves must have well-calibrated openings so that the slugs pass through readily, while the blanks remain on the sieve (Fig. 2-23). The separating may be effected manually or mechanically with vibrating screens. The importance of such a separating operation cannot be stressed enough. In fact, if small slugs (punchings from small holes, "French" stops, etc.) become attached to the blanks (lubricated and oiled stock creates favorable conditions for this kind of problem),

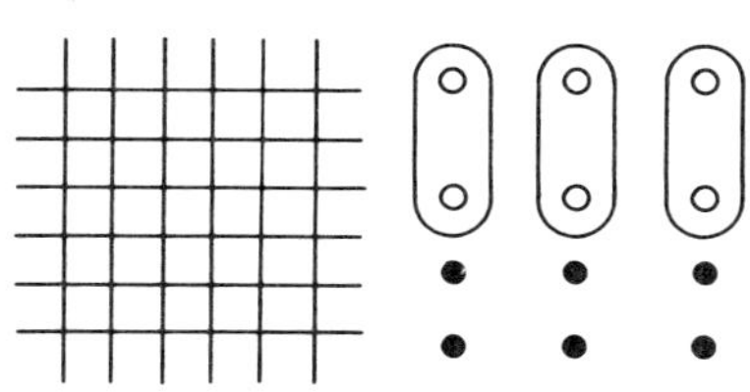

Fig. 2-23. Screen for separating stampings and small slugs.

then during additional processing (second operation punching, forming, bending, drawing, etc.) the slugs may easily damage the material and even the tool surfaces, causing spoiled work and even requiring repair work on the tools.

The trouble and expense of the extra separation operation may be avoided by following the simple solution shown in Fig. 2-24: Fasten a piece of sheet metal to the die bed to separate the blank opening from the slug opening and use two separate tote boxes. In multiplex dies this separation may not be possible because of the arrangement of the punches and die openings. In such cases, make the die-shoe somewhat thicker than usual and machine channels at two different levels (Fig. 2-25). Slugs and blanks will then fall from their separate channels by gravity. (It is assumed, of course, that an inclined press is being used.)

For handling of larger stampings (i.e., so heavy and/or large that one man cannot handle them alone, practically the same equipment is used as for sheets (see corresponding item above). In addition, trolleys and conveyors (belt, roller, overhead, etc.) are sometimes used.

In the case of delicate or important stampings of medium to large size, which are stacked between operations, it is good practice to put a reject stamping at the bottom of each stack, thus avoiding damaging a good stamping.

In extreme cases of very delicate surface finish and/or intricate shapes, it is recommended that a layer of paper, newspaper, or cardboard be placed between the stampings in order to avoid direct contact between them.

Note: Interoperational handling of stampings (from one press to the next press for subsequent operation, directly, i.e., progressive press processing) is considered as feeding and is dealt with in Chapter 10.

Storage of Stampings

This is a rather problematic activity, because storage of stampings is always temporary. You do not make stampings to keep, but to ship or to use in assemblies and

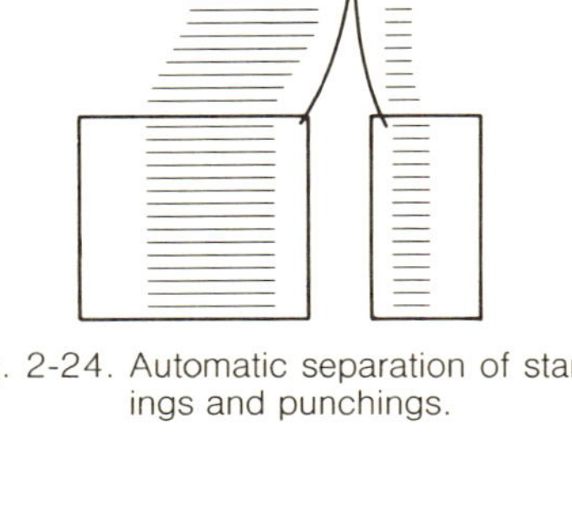

Fig. 2-24. Automatic separation of stampings and punchings.

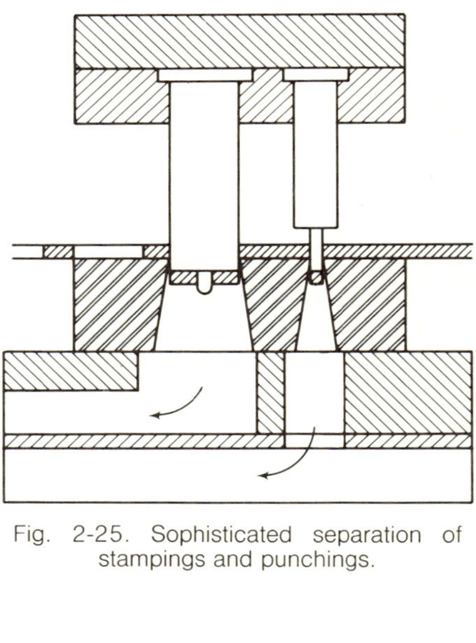

Fig. 2-25. Sophisticated separation of stampings and punchings.

then to ship. Consequently, it is not possible to assign specific areas and spaces for given product storage, because the contents of a typical storage room change continually, especially in the case of job shops. The best solution for such cases consists in assigning the different sector of the stamping storage room to roughly large groups of stamping types and trying to keep close track of the location of the single lots.

A very important rule is to keep the storage of finished stampings well separated from in-process stampings. The best method is to have sepatate rooms: keep at one end of the plant the finished stampings and at the sides of the plant the stampings between operations.

This subject of in-process storage is quite controversial. According to modern manufacturing methods, the process flow should be so scheduled that once a process starts, it should not be interrupted until the finished stampings are shipped, or at least stored. However, this is not always possible. Even in the best shops it quite often happens that there are interruptions in the process of stamping manufacture. In such cases—unless it is a question of a few hours or a maximum of one day—the parts in process should be taken from the manufacturing area to proper "in-process" storage rooms and kept there until needed.

Stored stampings (finished or between operations) should always be kept in proper bins, tote boxes, etc., except very large stampings, which may be located directly on pallets.

Scrap Disposal

The output of modern high-speed power presses equipped with properly designed dies and automatic feeding devices is so large that the disposal of scrap becomes quite a problem. If scrap is not removed efficiently, the production rate may be impaired seriously.

Scrap handling—which is equivalent to scrap removal and scrap disposal—is one of the "stepchildren" of metalworking manufacturing in general, not only of metal stamping production.

In many metal stamping shops, people do not attribute sufficient importance to scrap disposal. Strip skeletons and punchings are gathered in any convenient way and are sold to the scrap dealer; in some plants they are simply thrown away. This is wrong. If some time and study are spent in establishing proper means and methods for scrap collection and disposal, several economic advantages may be realized, such as savings in labor costs, increasing scrap sale value, and improving worker safety conditions.

Satisfactory scrap handling in the press shop may be performed either manually or automatically. In both cases punchings, slugs, scrap pieces, and other debris should not be allowed to become scattered on the floor around the press. Good housekeeping, safety, production efficiency, etc., are improved, if the scrap (possibly reduced to short pieces by means of scrap cutters) is dealt with correctly, according to a chosen handling method.

In the case of *manual* handling, the scrap pieces are collected in proper containers (barrels, loading bins, skid boxes, dump boxes with fixed or hinged bottoms, etc.) and transported in them to disposal areas. Scrap collection must be properly scheduled, several times per shift, at fixed time intervals.

In the case of *automatic* handling, if gravity is not satisfactory for direct scrap removal, then proper chutes and conveyors should be provided. These may be continuously moving or individual, movable or fixed conveyors for removing the scrap automatically from the interior of the presses and transporting the scrap material to a central scrap collection and disposal area. The general scrap collecting conveyors are usually either underground, aboveground, overhead, or combined (in medium-sized and large shops).

Figure 2-26 presents a Prab Scrap Processing System for metals, which includes a crusher, parts separator, centrifuge, and conveyors. Originally designed for machined chips, it is adaptable to metal stamping waste processing. (See also Fig. 11-8.)

Scrap skeletons are dealt with according to their original shape and feeding method. Thus, coil skeletons may be rolled up with scrap rewinding reels into coils, or may be chopped into short pieces with scrap cutters built into the dies or on the presses. Strip skeletons are handled the same as strips, and they may also be chopped by scrap cutters. Transportation efficiency is improved considerably by scrap choppers; in fact, the space requirements, i.e., volume of strip or coil skeletons, become enormously reduced when cut into short pieces. It has been found that the average ratio of reduction is three to one, or better. Thus the apparent capacity of transportation bins and the handling efficiency are increased almost in the same proportion.

Preference should be given to scrap cutters incorporated in the corresponding dies. Thus, it is convenient for the scrap chopper blades to be resharpened every time the stamping tool is reground. In addition, better timing of the chopping action

Fig. 2-26. Scrap disposal system. (*Courtesy of Prab Conveyors, Inc.*)

is ensured. (It is important to avoid cutting the strip skeleton during automatic feeding or piloting of the stock.)

All scrap (together with spoilt workpieces and rejects and material drop-offs due to improper sheet size, etc.) must always be held well separated according to type of metal and size, both for easier handling and for obtaining higher salvage values.

Excellent economy and efficiency are obtained by baling (compacting) the scrap, thus producing high density bundles of mill size, which are of much higher salvage value. Also, great space savings are realized in this way. Such baling presses—which compress the scrap from three sides—may be manual or semiautomatic (usually hydraulically powered) or totally automatic.

Rejected raw material must be put in special containers or on special pallets and a tag or a self-adhesive ticket attached with following data:

1. Source (name of supplier, warehouse, or mill)
2. Batch number and characteristics of stock
3. Cause of rejection
4. Date
5. Initials of inspector.

Such material should be collected and held in specially designated areas (near shipping and far from receiving) for easy handling and for avoiding using it, by mistake, for production. *Note:* Try to dispose of such material as soon as possible.

CHAPTER THREE

Press Tools and Their Care

Tool Design

Before starting the production of a new product or component the efforts directed toward improving press shop efficiency must begin in the tool design office. Experience shows that by applying the principles of maintainability and efficiency to the design of press tools, the production increase which results is usually much higher than that obtained by introducing new and expensive production equipment. The great advantages of high production offered by modern presses, with their attachments and accessories, may be achieved only if and when the design and construction of the corresponding press tools and the quality of their tool members are adequate.

Proper economical tools have to meet two chief goals: short production cycles and long tool life. While the first goal depends on correct design, the second goal depends on both design and construction.

The tool designer must determine the basic tool characteristics that affect productivity. There are several factors which must be carefully studied, with the principal ones being (1) production method; (2) feeding; and (3) ejecting (unloading). Feeding methods are covered in Chapter 10 and unloading methods in Chapter 11. Here will be discussed the production method which will answer the basic question: Are single operation tools or multiple-operation tools required?

Production Method

From the productivity standpoint, there are two basic details which must be taken into consideration: the difference between first operation and second operation tools and the difference between single operation and multioperation dies.

In first operation tools the stock is presented in the form of strips or coils (the latter offering the possibility for continuous, uninterrupted working). Second operation tools work with loose stock, unit stock, scrap, or blanks produced in a previous first operation: castings, forgings, machined workpieces, etc. Consequently, second operation tools can only work intermittently, unless some sophisticated automatic feeding device is provided.

The production rate of a first operation die is about 5–10 times as high as that of a second operation die. In addition, the hourly output of a press is practically the same with a single-action tool as with a multistage tool. These two facts strongly indicate giving absolute preference to dies which work with strips (or coils) and which perform more than one operation at each press stroke possibly all or nearly all the operations needed to transform a piece of sheet metal into a finished component.

The single-acting method, where every operation is performed with a separate die (usually on a separate press) successively, should be reserved for those cases where they belong, e.g., short run jobs or comparatively large stampings.

The enormous gains in production rate with a comparatively simple four-station plural die, in comparison with a set of four single-action dies, is shown in the following numerical example: Let us assume that the duration of every first operation is 1 second and the duration of every second operation is 7 seconds. So, in the case of the multiaction tool a stamping is produced every second, while with single-action dies the production of every stamping demands 22 seconds. (One first operation die means 1 second, and three second operation dies mean 3 x 7 = 21 seconds; total 22 seconds). Consequently, there is a ratio of 1 to 22 between the two methods.

The real gains in production efficiency are actually even higher. In fact, through elimination of the necessity of handling the semifinished products between operations, a great deal of time, labor, and expense are saved.

Additional important advantages of plural dies are:

Elimination of inspection between operations

Considerable increase in safety (reduction of accident hazards, see Chapter 13)

Only one press needed instead of several; thus better machine utilization, less floor space, less die setting time and labor, less supplementary costs

Quicker deliveries made possible by the increase in production rates

Reduced inventories; less storage space needed

Elimination of damage in interoperational handling and the chance of wrong loading

On the other hand, there are certain disadvantages in the adoption of multistage dies:

The cost (both first cost and maintenance costs) of a multiaction die is usually about 100% more than that of the total cost of the corresponding separate single-action tools

The size, shape, depth of draw, and other characteristics of the stamping may not be readily obtainable with multistage dies

The stock may not be suitable for production with multistage dies

More skills are required for setting up, operating, and maintaining multistage dies

The great variety of possible multiaction dies (with combined operations) may be classified into a few groups:

a. *Progressive dies* (follow dies) are the oldest and most commonly used type of multiple operation dies. Once used only for flat blanks with openings in them (two-stage punching + blanking dies were the usual case), today, they are also employed for various kinds of formed stampings, with several types of press operations, such as drawing + blanking; punching + notching + forming + blanking, etc. Other press operations which may be performed with progressive dies are: flaring (of holes), countersinking (of holes), extruding, lancing, trimming, embossing, swaging, and tapping (of holes).

The number of stages (stations) in a progressive type die cannot be too large because of the much higher tool and maintenance costs and greater difficulties of operation. On the other hand, it is not good practice to crowd too much action into one stage. In cases where severe forming is needed, it is preferable to increase the number of stations, dividing the forming into several, successive light phases.

b. In a standard progressive-type die, the precut stamping is usually carried along with the original metal strip or coil, until the final (blanking) station. There is a variant of this procedure, the *cut-and-carry* die. Here the surrounding strip is almost completely trimmed away, leaving the stampings connected by a small bridge of stock. Thus they are carried from station to station until the final station, where they are severed.

c. *Transfer dies* are also essentially progressive dies, because they perform successively several press operations until the completion of the components, passing them along automatically from station to station. (Also, with transfer dies, at each ram stroke a finished component comes out of the press.) However, there are a few basic differences between standard progressive dies and transfer dies, which have the following characteristics:

1. They work with special multiplunger presses ("eyelet" presses), each operation being performed by a separate die, individually actuated (though simultaneously with the other dies). The plungers are usually arranged in a straight line; however, there are also transfer presses where the plungers are arranged circularly, i.e., around the periphery of a round table (similarly to dial feeding).
2. The stampings are completely severed in the first station; i.e., they are not held by the strip skeleton, but are carried along from station to station by transfer mechanisms (mechanical gripping devices).
3. It is not always necessary to start with strip or coil material; the transfer press also may be fed with previously processed blanks or preformed stampings.

Contrary to common belief, transfer presses are efficient and economical not only for very large volume production, but also for comparatively moderate, medium size batch productions.

d. *Compound dies* are telescope-acting tools for simultaneous action (or, at any rate, for actuation in one ram stroke) of usually two, but sometimes three or even four, press operations, such as piercing + blanking, blanking + drawing, blanking + drawing + punching, etc.

Their chief advantages are the relationship between blank contour and punched apertures is held with maximum accuracy; flatness of stampings is ensured (if they are not formed stampings); a smaller overall size than for progressive dies is possible for the same components; longer die life; and smaller amount of tool-room equipment is needed. Their limitations are higher first costs than any other kind of tooling; slower operation due to retarded ejection of the finished stampings; and difficulty of automatic stacking of stampings.

Not only cutting and drawing dies but also forming dies are sometimes made in a compound design. In such cases more or less complicated dies are built (incorporating solid, sectional, sliding, rotating, oscillating, cam-actuated, etc., members) which perform the necessary partial bending operations (phases of the total forming process) successively until the completion of the stamping.

e. *Combination dies*. There seems to be no general consensus about the right meaning of this expression. Some people use it instead of compound dies, others for any multiaction die. This author uses it to denote progressive dies having a final stage which is "compound," i.e., where simultaneous piercing and blanking take place.

f. *Multiple dies* perform the same or the same type of operation simultaneously two, three, or more times. A progressive, compound, or combination die may at the same time also be multiplex, thus attaining very high production rates. Also, the simultaneously produced stampings may be of different shapes in order to minimize scrap. Other characteristic cases are those where in a given product a right-hand and a left-hand component are needed (duplex tools). In all these cases multiple dies offer the additional advantage of a better stock utilization factor. Sometimes also second operation punching dies are made multiple in order to save time, increase accuracy, and increase productivity.

Press Tool Maintenance

Introduction

Keeping dies in proper operating condition is of the utmost importance from the production economy standpoint. Since, usually, there is only one tool for a given, particular case (in contrast to machining), when a die is not in operating condition, production must be stopped until a replacement can be found or the original die repaired. Thus it may happen that, because of a die which is out of order, workers may be idled for days or even for weeks. Furthermore, with modern high-production presses, a shutdown for even a few hours means a serious loss of production.

It is an established fact that a well-organized tool maintenance program will make a substantial contribution to overall higher-production efficiency and, consequently, will lower unit costs.

Planning a sound tool maintenance control is based on past experience in actual production. Therefore, it is important to keep maintenance logs of every tool, where performance data ("batting averages" = number of successful hits divided by the number of setups), causes and solutions of problems, etc., are recorded correctly.

Significance of Tool Maintenance

In spite of its importance, the maintenance of press tools does not always receive the attention it deserves. (Tool maintenance activity is considered as non-productive and therefore it is often given a low priority in the press shop.) With the exception of a few large companies, most people connected with pressworking of metals do not recognize the significance of die maintenance, and so data available on the subject are scanty. Any toolroom job that is not invoiced to customers (i.e., direct labor cost) is simply charged to overhead. However, by piecing together scattered reports, data from the author's own experience and the results of a personal small-scale survey, the author reached certain conclusions which are summarized here.

The ratio between production (running) hours and servicing (maintenance) hours varies between wide limits. The chief determining factors are the plant, the press, the tool, the stamping (its type, size, shape, etc.), the characteristics of the stock, toolroom equipment, quality of the press operators, quality of the toolmakers, etc. At any rate it seems that the following data pretty well cover the subject:

a. For very favorable cases of simple, thin, flat stampings of small overall size, produced in single-operation tools, the ratio between production hours and maintenance hours may be about 100:5 to 100:10, i.e., a few minutes of maintenance time for every hour of production

b. With multistation progressive dies, the ratio is between 100:15 and 100:30

c. Complicated dies (including forming stations, cam-actuated slides, and similar items) can require ratios of 100:40 to 100:60

d. Finally, in extreme cases (e.g. lamination dies of the progressive type, working with very abrasive silicon steel sheets), the ratio may be 100:150 (i.e., more maintenance time than actual production time)

Maintenance costs for different types of press tools range from highest to lowest in the following order: (1) forming and drawing tools; (2) progressive and compound dies; (3) single-operation punching and blanking tools. Repair costs per hour to a die must be assumed to be higher than the cost of building a new die, per hour. In fact, the *total cost* of die maintenance, during the whole useful life of a mass-production tool, is almost always a multiple of the original cost of the tool. In fact, the average ratio is 2:1 for cutting tools and 3:1 for forming and drawing tools.

Consequently, it is a serious mistake when evaluating the cost of a press tool to take into account only its initial cost since it is only part—and a smaller part—of the total cost of a tool. A proper estimate would figure both the direct costs of die maintenance and the indirect costs of press downtime for repairs and servicing of the tool. For high-production tools (with a long life expectancy), it is only practical to spend more on the building of the tool, if in this way, the die requires less servicing and maintenance. The initial extra expense will always be a fraction of later expenses in time, labor, and production. A higher-quality die will require less maintenance than a lower-quality tool.

Scope of Die Maintenance

Press tools vary widely in type, size, weight, difficulty in handling and operating, and frequency of use. In addition, every shop is different and has its own peculiar characteristics, such as per space distribution, equipment, personnel, tradition, history (development), business trend, etc. So it is impossible to give a definitive set of rules to follow for the maintenance of dies; only a skeleton outline can be provided, which must be adapted to the individual needs of the shop.

The die maintenance activity includes the following items: tool administration, tool handling (transportation, storage, and inspection), and tool servicing or reconditioning (resharpening, repolishing, repairs, replacements, etc.), in addition to taking care of certain tool design details.

Tool Administration

In order to be able to keep track of tools and to know everything about their life, it is necessary to have well-kept files at one's disposal. A well-managed card index for tool control must be maintained in order to chart die movements at all times and to pinpoint a die's location instantly. To set up such a system, every tool should be given an identification number, stamped both on the die shoes and on the punch holders. A suggested method is to mark (with steel stamps), on both parts of the die set, the part number and the sequential operation number. If there is more than one die for the same purpose (standby tools or replacement tools), stamp roman numerals (I, II, III) after the part number (Arabic numbers). Universal tools should be set off by a suffix, e.g., "U."

In addition, it is common practice—and a very good one—to paint on their storage location symbols (e.g., the coordinates of rack numbers and shelf letters). Finally, some shops use differently colored painted marks (or combinations of colors) to identify the tools that belong to the same category.

Handling of Dies

Dies are often mishandled. An unmounted blanking punch may be dropped edgewise upon the cutting aperture of the corresponding die plate, damaging cutting edges (Fig. 3-1), or a punch may be forced too deeply into the die opening, causing serious harm which must be repaired before the tool can be put to work. Safe handling of dies is best effected by means of bumper blocks put between the stationary and movable members of die sets, as well as the stationary and movable parts of press tools not mounted in die sets. At the very least, heavy cardboard or soft metal sheet should be placed between the cutting or forming members of dies during handling—transportation and/or storage.

At the risk of belaboring the obvious: Dies must be held in racks, on trucks, etc., in their natural or working position; never turned on their sides nor upside down. They should never be piled one upon another, either in storage or during transportation.

At the completion of a production run, when a die is removed from the press, it should be a standing rule (applied and enforced constantly) to keep with it samples of the last stampings produced and part of the last stretch of strip skeleton. This serves two purposes: First, die inspection prior to storage is greatly facilitated; second, when the die is to be used again, die setting becomes easier, quicker, and more accurate if a finished sample is at hand. Compound tools with delicate, structurally weak tool members are somewhat protected against overloading in storage if a finished stamping is placed in the tool between male and female stamps.

If a die is equipped with some loose accessory (e.g., alternate locators), it should be tied to the die after removal from the press, otherwise there is a danger of losing it during handling. A better solution—whenever possible and practical—is to attach those tool members permanently with small chains, cables, etc.

If a die has been in storage for a long time, it should be taken apart, cleaned, and lubricated before it is delivered to the press room for production. Tool checking should be a rule, even with dies stored for short periods of time.

Storage procedures should permit only storage room attendants in the die storage

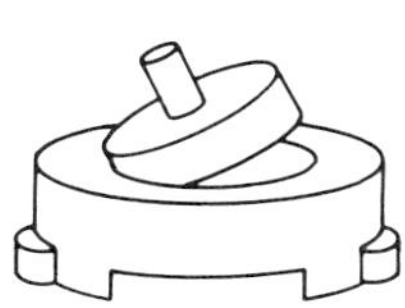

Fig. 3-1. Incorrect storing method for open dies.

area. Two benches should be provided outside the storage room: one for "shipping" and one for "receiving." The first one should be reserved for dies to be set up the same day or early the next day, the second one should be for dies removed from production.

Inspection before Storing

Ideally, no die should be stored if it is not in perfect working condition, so that it may be delivered to a press at a moment's notice. (The exception would be a die which has become obsolete, but not yet to be definitely discarded.) Therefore, before storing, a die should be thoroughly cleaned (all dirt and oil removed) and then checked carefully for any defects by an experienced toolmaker. The checking should always include the last samples from the production run, the last end of the strip-skeleton, and possibly also the last slugs or punchings. This facilitates (through checking the cut edge conditions, burr formation, dimensions, contours, finishes, possible distortions, etc.) evaluating the conditions of the cutting and forming edges of the tool members. In addition—in order to increase the efficiency of the checking activity—the tool should be accompanied, if possible by gages (if used by the operator or diesetter, routinely) and an up-to-date blueprint of the part (with any written specifications).

For dies in obvious need of reconditioning or repair, the tool should be accompanied also by a piece of raw material (a strip or some blanks) for tryout after servicing. Of course, in such cases where trouble has been experienced, it is compulsory to submit a detailed written report containing a description of the problem encountered and suggestions for reconditioning procedures. (This report should be presented by the diesetter or the press shop superintendent.)

In the case of long production runs, die inspection should be performed periodically during production (about once a week or so) and, if possible, without dismounting the tool from the press.

Furthermore, during regular operations with the die, routine quality control checks should be performed by the inspection department to determine whether or not the product meets specifications. If not, the press is stopped, and the die is sent to the repair shop (unless the damage is so slight that the tool can be repaired in the press).

Pressdown Reports

It is sound practice to have the diesetter or the press superintendent send his report to the engineering department along with the die. This report should include suggestions for any more or less substantial modifications of the tool in question, if needed. If such a procedure is consistently practiced, considerable advantages may be achieved. Of course, such reports should not be simply filled out forms which are filed away. They should be properly conveyed to the tool designers (who must sign the reports before these can be filed away) in order that they can apply what actual experience teaches in future designs. In fact, toolroom and tool design office can learn about performance of a tool only through reports of those people who actually use the tools when they are operating. Such reports might also cover the presses, attachments, feeding and safety devices.

Transportation of Tools

Small tools are transported by means of hand-driven carts. It is strongly recommended that such carts be provided with rubber tired wheels in order to avoid problems created by strong vibrations during transportation. Such carts should not be overloaded with too many and/or too heavy tools. Never put one tool on top of another; it is always better to make an extra trip. Medium-sized tools are usually transported by self-propelled fork lifts and large tools by means of cranes.

Die Storage

The storage of dies does not always receive due care and consideration with consequent reduction in production efficiency.

Dies, when not in use, must be kept together in orderly fashion, safely and efficiently, in a space expressly reserved for this purpose. The size, shape, location, and organization of die storage varies widely among shops, according to their type, production schedules, tool sizes, etc. In small plants, usually a corner (or a wall) of the press shop (hall) is arranged as a die crib. In larger shops, a separate room may be set aside (or a couple of rooms, if necessary) in the same building, adjacent to the press shop or to the toolroom. Even cellars, mezzanines, or balconies may be utilized successfully as storage rooms. In very large plants, dies may even be held in separate buildings of their own.

Die storage rooms should be reasonably clean and, above all, dry, in order to prevent oxidation of the tools. Both the floor and the racks must be strong enough to bear the anticipated loads (with ample margins of safety). Under no circumstances should a die be left directly on the floor. Dust is one of the greatest enemies of press tools.

Small dies (which can be lifted easily by one man, i.e., a maximum of 30 pounds or so) are best kept in steel or wooden dust- and humidity-proof cabinets. The ever-present dust in the shop, when mixed with lubricants, forms a powerful abrasive that slowly destroys guiding and active surfaces of tool members.

Medium-sized tools, which are handled by mechanical means, are kept on shelves in open racks made from structural steel, wood, or mixed materials (e.g. steel frame and wooden planks). The height of the racks and shelves must be sufficient so that storage and removal of tools can be effected easily and quickly. Free openings of 8, 10, 12, and 16 inches have been found satisfactory in several shops.

For medium-sized and large tools it is good practice to provide supporting rails on the shelves so that fork lifts can easily reach below the dies to pick them up.

Skid platforms are excellent auxiliary means for handling large dies. Very bulky and heavy dies must be stored at floor level, on wooden or steel rails, in order to avoid direct contact with the floor and to be picked up easily by fork lifts. Often such tools are stored in the open, in which case they must be protected from the elements either by means of waterproof covers or by proper coatings (e.g., easily removable grease, soluble in solvent). Even if the dies are stored indoors, they should be coated lightly with mineral oil, in order to prevent them from rusting, or they should be wrapped in rust-preventing paper (or cellophane, waxed cheese-cloth, plastics, etc.), sacks, or bags. These also act as a protection against dirt and dust.

Whatever their size, the tools that are used more frequently should be kept in an easily accessible location. In small shops, satisfactory results can be obtained with a mixed storage system, which consists of reserving a separate area in a handy location for the production people for tools that must be used frequently (if possible, in the pressroom itself). Other tools may be kept in storage rooms at greater distances. In general, storage areas for dies should be centrally located with respect to the presses and toolroom in order to keep handling and transportation to a minimum.

Dies may be arranged in storage according to: (a) type of operation (cutting, forming, drawing, etc.); (b) size; (c) chronological order (of construction); (d) grouped together according to end product or part. The most suitable system must be determined individually in each case.

Whatever arrangement is chosen, it is necessary to assign a regular place to every die so that it is quickly located and easily accessible with a minimum of shifting of adjacent tools.

When a new die storage room is being devised, do not forget to·take in account possible (and probable) future needs (in order to avoid the storage rooms' ever-present evil: high-density storage of tools). A practical piece of advice is to double the present available storage space, so future expansion is amply taken care of.

Servicing or Reconditioning Operations

The inspection of a tool, prior to storage, may detect any of the following defects (whose elimination is a prime objective of servicing): (a) evidence of natural wear; (b) consequences of careless operations or sloppy setups; or (c) breakdowns (mechanical short-circuits, material fatigue, accidents).

Routine damage in cutting dies includes: dulling of cutting edges, deformed or worn stop pins, worn stock gages, worn pilot pins (due to abrasion), trouble with slugs jamming, and magnetized tool members. In forming dies, it is chiefly scoring and wear of the forming (drawing, bending) edges, and, in general, of the active working surfaces. In any type of die there may be found worn or bent ejectors, scored or worn die-set leader pins and bushings, fatigued springs, etc.

Die breakdowns resulting from carelessness would include miscuts (due to poor feeding, improper gaging, heavy burrs, poor stripping, incorrect piloting, poor material guiding, press retripping, etc.); partial cuts; double blanking (or double punching); double forming (double header); trouble with slugs, press misalignment; die misalignment; press overloading; wrong quality of stock; wrong size of stock (thickness and/or width); loosening of some tool member (by vibration caused by the impacts); improper stock lubrication; caked oil blocking ejectors; improper tool setting; and faulty nesting (in second operation tools).

While a breakdown may be caused by practically any member of a die, it is more usually due to those parts which are heavily stressed and yet are comparatively weak. This would include punch breakage due to slenderness; chipping or breakage of cutting edges of punches and/or die plates; spring failures, etc.

Reconditioning

This includes such activities as repairing and polishing scored surfaces; replacing worn, chipped, or broken tool members; resharpening dull cutting edges; and demagnetizing tools and tool members.

Usually, these jobs are performed in the toolroom. However, if the size of the plant justifies it, it is worthwhile to install a separate repair and maintenance tool-

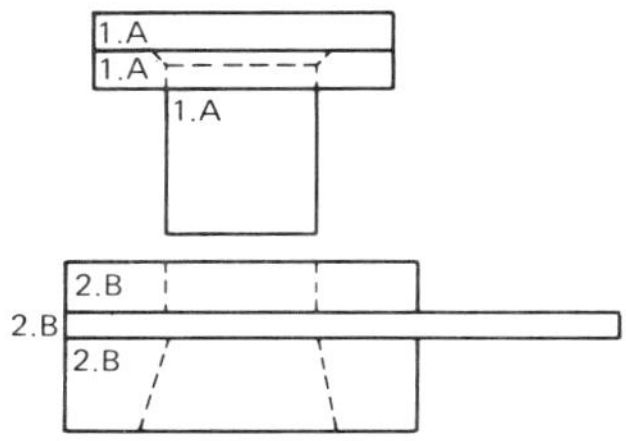

Fig. 3-2. Marking die components according to their positional relationship.

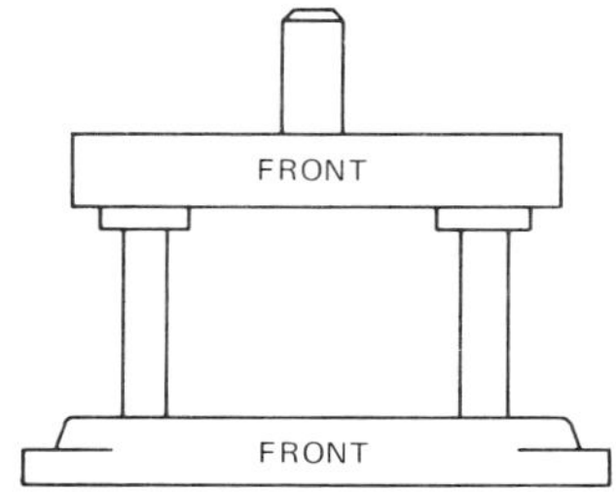

Fig. 3-3. Mark the front side of die holder and die shoe of die sets.

room adjacent to the storage room. This latter practice is to be encouraged because of the substantial advantages offered. It should be understood that the toolmaking department is always responsible for *both* the building of new tools and the maintenance of existing tools.

The following check list for press tool servicing has been found very helpful in actual practice.

1. Before disassembling the die, mark the parts, if necessary, in order to get them back into their proper place and relationship. They can be marked with numbers or letters or punchmarks (Fig. 3-2). In this way, it is easy to prevent wrong reassembly after servicing. In addition, it also is good practice to mark the die set in a very conspicuous manner indicating the correct orientation, in order to ensure proper operation (Fig. 3-3).
2. Take the die apart and check for irregularities.
3. Examine every part, especially the active tool members, i.e., punches, die plates, stamps, etc.
4. Appraise the condition and quality of the die and of every tool component; decide what has to be done to each.
5. Sharpen dull punches and die plates.
6. Replace and/or repair worn or damaged tool members such as springs, dowel pins, ejectors, stops, screws, bolts, or die set.
7. Perform any other operation, according to what was decided in item 4.
8. Assemble the die in proper sequence but before assembling carefully clean every tool component, and remove all chips, burrs, dirt, and foreign matter from spring pockets and from any blind hole or recess.
9. Make nine or ten samples.
10. Submit samples for inspection, after you have satisfied yourself that the samples are all right.
11. After passing inspection, store the die or return it to press shop.

Resharpening of Dies (see also Chapter 15)

The cutting edges of punches and die plates, after a number of operations, become dull and must be restored by grinding. The need to resharpening the active tool members is usually indicated by the height of the burrs on the stampings, slugs, and strip-skeleton-opening edges. When the burr height exceeds the specific allowance limits (which, in turn, depend on the function of the stamping and are different for each type of component and application), the punch and the die plate must be reground.

In some shops the tools are resharpened after every production run, before they are stored, or at predetermined intervals, after a fixed number of hits.

Here follow a few practical hints for resharpening cutting dies:

Before setting up a tool member in the grinder, carefully clean it, removing slugs, chips, blanks, etc. from die plate openings.

Except in rare cases, grinding should be performed wet (with appropriate cooling liquids). Among its other advantages, a better finish is always obtained with wet grinding.

Use only light cuts, with the proper wheel. Heavy cuts "burn" the tool steel, ruining it beyond any possibility of repair.

A light stoning of the cutting edges after grinding prolongs the useful life of tools.

In multioperational dies (progressive, multiplex, etc.) all the cutting punches and die plate sections should be resharpened at once, even if some of them are still not quite dull.

When sharpening small perforating punches, prevent their bending and breaking by supporting them with the stripper plate. Place the punches in their corresponding punch plates and guide holes in the stripper, letting them protrude about one-eighth inch or so (Fig. 3-4). With very small punches, where the cutting point is smaller than the punch body (shouldered punches), support the cutting points further by pushing over the punch ends a piece of the corresponding strip skeleton, which has all the punched and blanked openings in it (Fig. 3-5).

If a punch or a die plate has "shear" (inclined cutting edges), this fact and the amount of shear should be posted on the tool card so that the toolmaker is alerted to the special grinding method required.

In grinders equipped with magnetic chucks (which is the usual case), the punches and the die plates may become magnetized and retain part of the magnetism even after reversing the current of the chuck. Then, when the tool is cutting steel sheet, small chips and slivers will adhere to the cutting edges

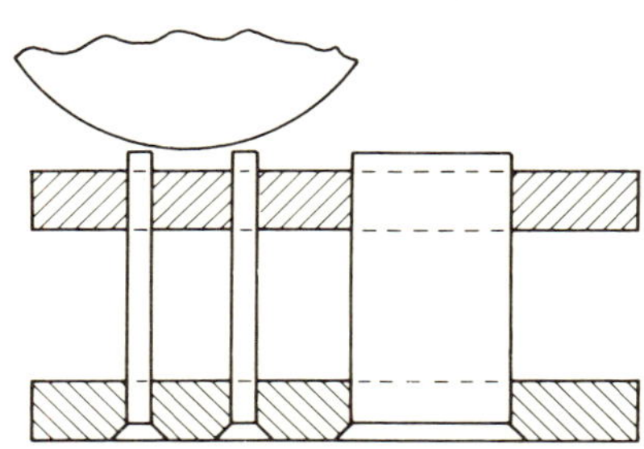

Fig. 3-4. When sharpening small diameter punches, support them with the stripper plate.

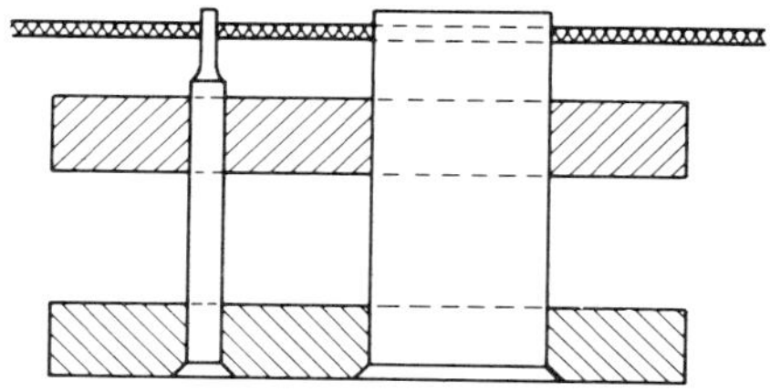

Fig. 3-5. In case of frail punch points, in addition to the stripper plate, also use a piece of punched strip skeleton to support the punch points.

and damage them. Therefore, such tool members should receive additional demagnetization before assembly and then be checked to determine if they retain any magnetism. Such magnetization may result simply from the interaction of the earth's magnetic field and the vibration of the tool as it operates.

After sharpening, clean all die members thoroughly before assembly. Remove all sludge, grit, and other foreign matter.

Cutting dies should be designed so that resharpening is possible without having to remove either the leader pins or the tool members from the die set. (See item "Tool Design" below.)

Repolishing

As with cutting dies, drawing, bending, and other forming tools also must be reconditioned as needed. The exact time for this depends on such variables as the stock thickness, the shell or stamping dimensions and characteristics, the lubricants, the press, and the particular tool, but it is indicated whenever surface imperfections (scratches, nicks, scorings) appear on the stampings.

Restoring

Restoring of worn tool members depends on the individual characteristics in each case. Cutting edges may be restored in many cases by hard-facing or by substituting inserts (or sections) in sectional dies; in forming tools hard chromium plating and/or special heat treatments are used.

Tool Improvements

If the design of the tool can and must be improved at all, the best time to do it is while the tool is being repaired.

Standby Tool Members

A rundown of standard die parts to be kept in stock would include:

a. Universal parts: screws, bolts, nuts, dowel pins, perforating punches (or, at least, blanks for them); pilot pins (or blanks); die bushings; stripper plates and punch plates (without holes); safety microswitches; stop pins; stock guides; and springs.

b. Individual parts: critical components, such as frail, replaceable inserts or segments of sectional dies, and special punches and die-bushings of high-production tools.

This standby principle is sometimes extended to complete tools (and equipment parts) where production schedules are very critical and tight. (More about this subject under "Standardization" item.)

The author was once involved in an actual case (in Chicago) of a supercritical product schedule where the standby principle was applied to a whole press line (made up of small punch presses). Every operating press had a fully equipped (and tooled) identical standby press placed only a few feet away. Whenever something happened to an operating press, its production was taken over in a few minutes by its standby press, and so the downtime was held to an insignificantly low value.

If the correct replacement part is not available for a damaged critical tool component, in an emergency a part may be removed from another tool which is idle. However, such a procedure should be resorted to only in exceptional cases, and the part should be replaced as soon as possible.

Final Inspection

As soon as a tool is repaired, it is inspected (together with the samples submitted by the repair toolmaker who tried out the tool) in order to see whether or not everything is satisfactory. After proper lubrication and perhaps coating with rust preventative, the tool goes to storage or back to the press shop.

Appendix: Tool Design

The tool designer may contribute considerably to the improvement of the tool maintenance conditions. Many problems may be avoided during press operations and servicing if the tool designer keeps in mind what trouble may arise if certain tool details are not designed carefully enough.

This subject is a vast one and whole books have been written about it. Here are mentioned only a few practical suggestions for making tool servicing easier, quicker, safer, and less expensive.

Generalities

Tool members that need a great deal of attention (care, repair, and maintenance because of intricate sections, frail components, etc.) should be so designed that replacement is easy and quick, especially in the case of small cutting shapes where there is a great danger of punch breakage due to high compression stresses.

For those cases where small punches must be frequently changed or substituted, quick-change punch-holding methods should be made possible.

Tool Handling

For handling heavy dies and die sets, it is good practice to provide the bottom members (die holder, die shoe, etc.) with through-canals. In these slots pry bars, chains, or rope can be easily inserted for lifting and transporting the tool (Fig. 3-6).

For easy and safe handling of dies—especially if they are heavy—in the strippers, pads, and other large members, proper tapped holes for eye-bolts or hooks (Figs. 3-7 and 3-8), or smooth holes for lifting lugs, pins, or rods (Fig. 3-9) should be provided. Such holes should be provided also in the top, front, and rear surfaces of large dies (Fig. 3-10 where suggested proportions are shown).

In case of medium- to large-sized dies it is good practice to provide straps or bolts to hold together the movable and stationary parts of tools not mounted in die sets, in order to prevent them from falling apart during transportation (Fig. 3-11).

If a tool has some protruding details (e.g., ejector pins) it is recommended that they be made readily detachable for easier and more practical storage.

Sharpening

Mark the brand name on hardened tool steel components (chiefly punches and die plates) that are to be reground routinely; such information is a great help when sharpening the parts. Obviously, the marking must be located on such spots that are not ground away (e.g., the sides).

Provide sufficient sharpening allowance on punch and die plate heights.

All protruding accessories (stop pins, shedder pins, stripper plates, pilots, locators, spring plungers, etc.) must be easily removable for grinding.

When in a progressive die there are combined cutting and forming stations, the forming stamps should be backed up by filler blocks that can be ground off the same amount as the cutting members, in this way preserving the correct height relationship among the different tool components (Fig. 3-12).

Cutting (blanking and perforating) punches cut only with their outer edges; therefore, it is convenient to relieve their bottom center, leaving only a narrow outer band. In this way, the resharpening operation is reduced considerably. Round punches (Fig. 3-13) and nonround punches (Fig. 3-14) may be processed in this way.

Figure 3-15 presents a very simple, yet little used practical detail: Putting two die-set posts (of a four-post die set) in the punch holder plate and two in the die holder plate. A small surface grinder can then be used for sharpening the cutting members of the tool, without having to dismount them from the die-set.

It is also good practice to take in account maintenance difficulties in the case of forming dies. Those sides that are subject to wear should be made in sections and precisely so that they are easy to regrind. This is achieved best if the sections are so designed that the sides that are to be reground have parallel countersides. (Fig. 3-16). In such a way, it is a simple matter to regrind by putting them on the magnetic chuck of a plane surface grinder.

Miscellaneous Tool Details for Easier Tool Servicing

Select a die set with easily replaceable guideposts and bushings or with ball-bearing guides.

For punching small holes, use die bushing inserts instead of drilled (and reamed) holes in solid die plates.

In progressive dies, use separate units for each stage.

Often it is very convenient to provide screwdriver separating notches in solid strippers, punch pads, die blocks, etc., for easy and quick disassembly of tools (Fig. 3-17).

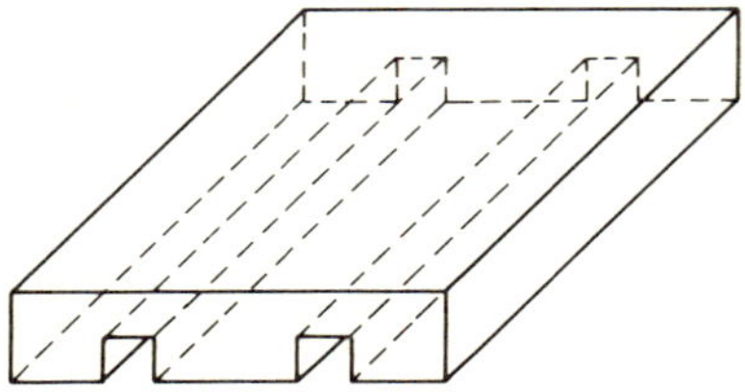

Fig. 3-6. Chain slots at the bottom of die shoes are a great help in tool handling.

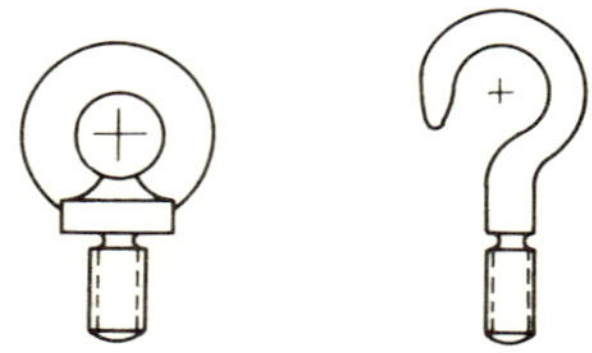

Figs. 3-7 and 3-8. Eye bolts and hook bolts efficiently aid in handling of heavy and large dies.

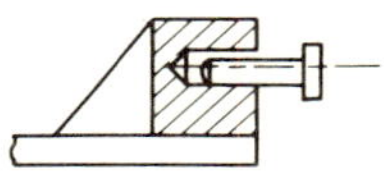

Fig. 3-9. Smooth pins put in side holes facilitate die handling.

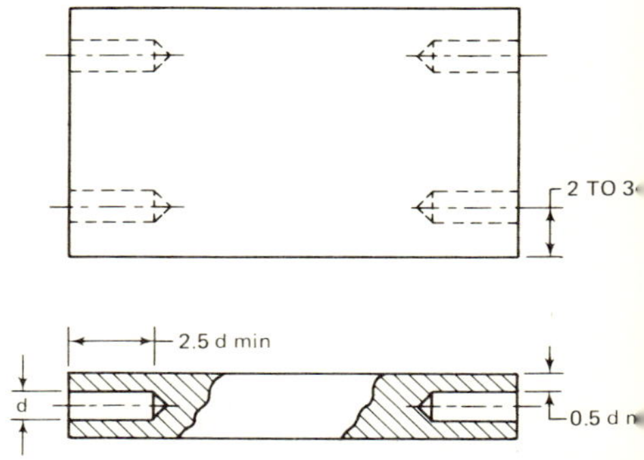

Fig. 3-10. Suggested dimensions for side holes in heavy plates.

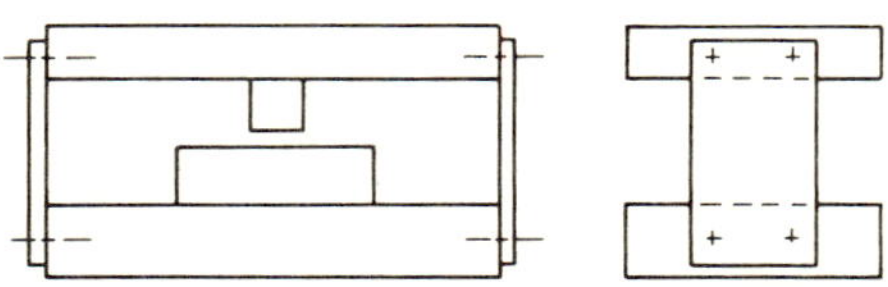

Fig. 3-11. Flat steel straps for safe handling of open dies.

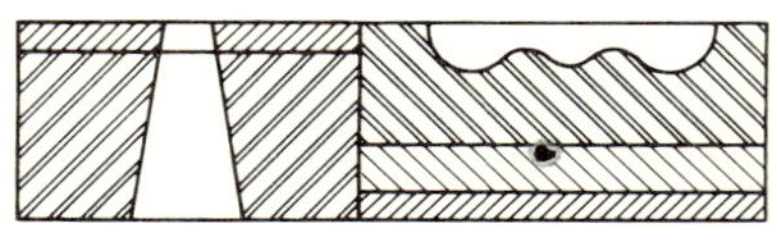

Fig. 3-12. In progressive dies with forming station(s), provide spacer blocks below the forming stamps, and, when sharpening the cutting sections, also grind the spacers.

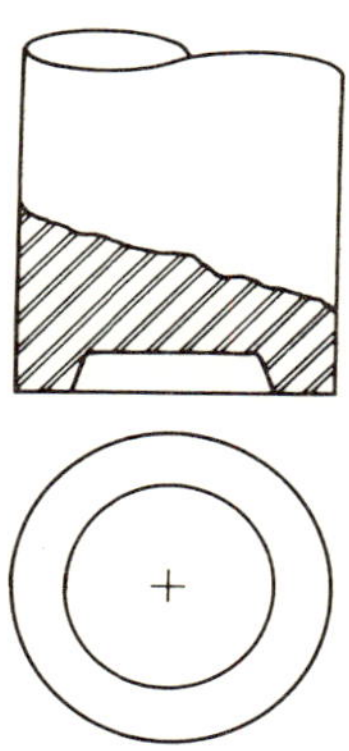

Fig. 3-13. Relieve the bottom center of round punches for easy sharpening.

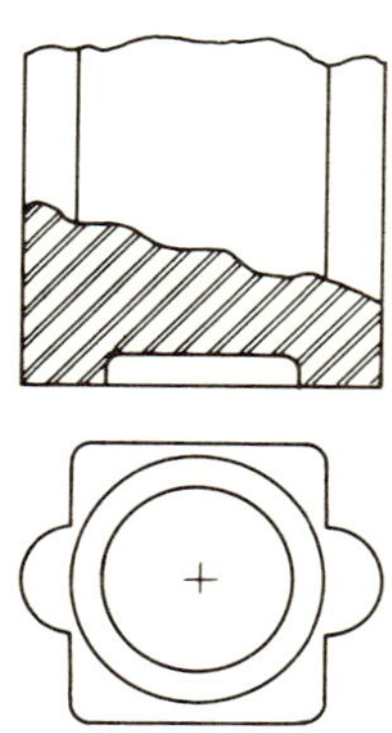

Fig. 3-14. Nonround punches may have relieved bottoms to facilitate sharpening.

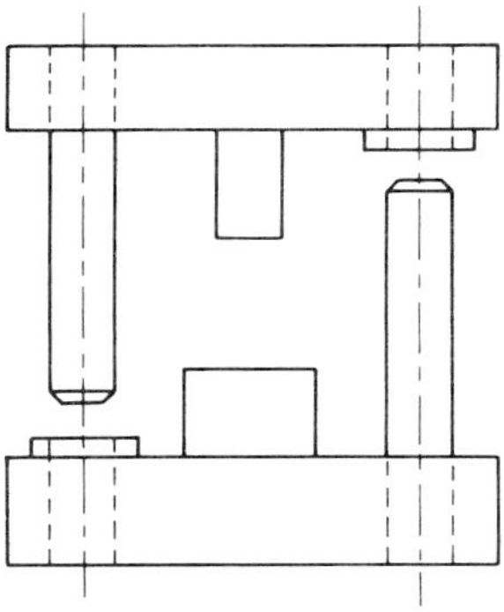

Fig. 3-15. For ease and rapidity of resharpening of dies, use die sets with properly located leader pins and bushings: one side up and one side down.

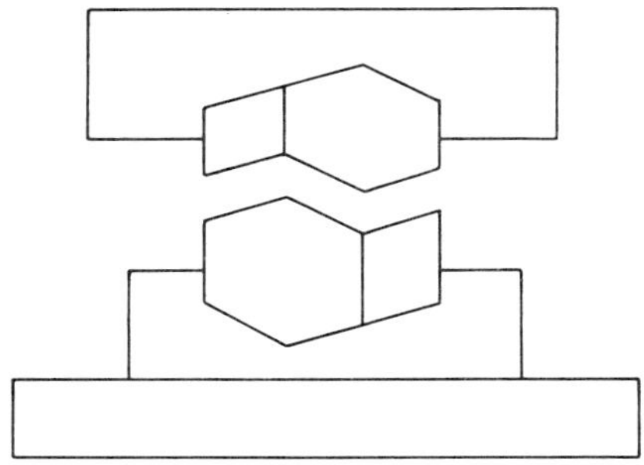

Fig. 3-16. Special sectional forming stamps should have parallel sides, so resharpening is made much easier and more accurate.

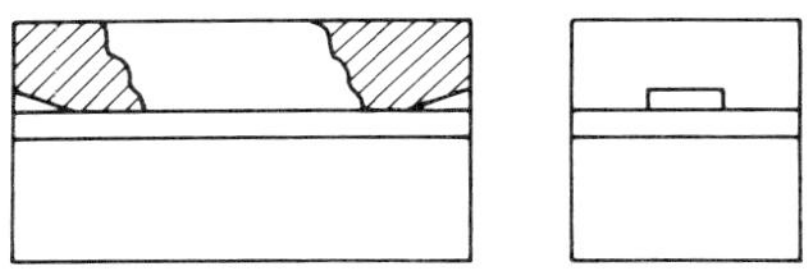

Fig. 3-17. Separating notches in strippers, pads, etc., facilitate disassembly of dies.

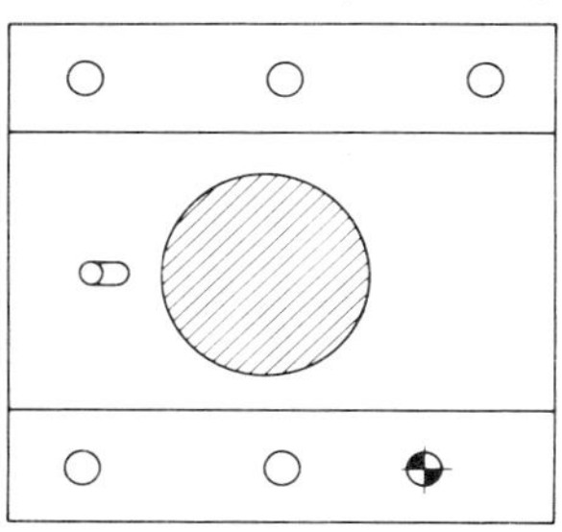

Fig. 3-18. By locating one dowel pin off-center, incorrect assembly is prevented.

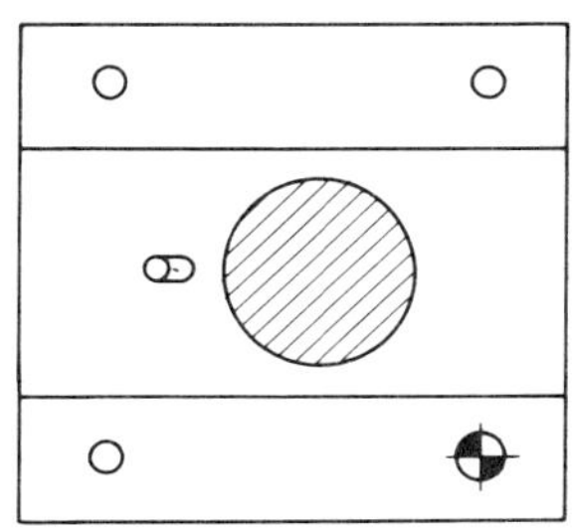

Fig. 3-19. In case of small dies, select one dowel pin larger than the others, in order to prevent incorrect assembly of the dies.

Symmetrical dies are liable to be assembled in reverse with possible disastrous consequences, therefore, they should have a foolproof design. In order to avoid an incorrect assembly for such tools, use either of the following:

a. In large dies locate one of the dowel pins off center (Fig. 3-18).

b. In small dies make one of the dowel pins of greater diameter than the others (Fig. 3-19).

Standardization

Because of their individual character, dies seemingly offer little possibilities for standardization. However, if we take a closer look and analyze the situation more thoroughly, it will be found that in these special tools there are only a few parts where the design depends directly on the individual, specific component to be produced (or operation to be performed); the other tool members work indirectly by holding or guiding the direct acting tool members and/or the stock; therefore, they readily lend themselves to partial or, in many cases, even total standardization. Also, direct acting tool members often can be standardized, for example, small round perforating punches and die buttons, as well as other items.

Through standardization we obtain true and easy interchangeability of certain die members. This fact ensures quick replacement of faulty tool members. In case of mechanical short-circuits, wear, etc., a standardized tool member is quickly, easily, and inexpensively replaced, because interchangeable spare parts (held in readiness in the toolroom store) always fit. Consequently, there is reduced downtime of the press and production efficiency is improved.

Classification of Standards for Die Members

There are three major types of standards that may be employed successfully in the designing and building of dies: (a) national standards; (b) suppliers' standards; (c) company standards.

National Standards

There are comparatively few of these—mostly in the nature of punches and die buttons. The American National Standards Institute (ANSI) and the National Association of Punch Manufacturers (NAPM) have issued some. It seems, however, that there is no great activity in this field owing to the outstanding job done by individual tool component manufacturers.

Suppliers' Standards

There are several manufacturers who specialize in producing, on a commercial basis, a large variety of common tool and die members. Owing to competition and mass production, their selling prices are usually below what would be the cost of production if the individual toolmaker of an independent toolroom were to make them singly with less adequate means and methods. Besides, the quality of commercial products is usually higher. Consequently, it is, from every standpoint, advantageous for the die designer to select one (or a few) supplier(s) for all the available ready-made tool members and adopt their standards insofar as they apply to the dies they build.

Tool members that can be purchased from suppliers form an amazingly long list. Here follows a listing which has no pretension of being complete: whole piercing or punching units; standard and special complete die sets; die holders (die shoes); punch holders; guide pin bushings (guidepost bushings); guide pins (pillars, leader pins, guideposts); floating adapter shanks; separator rings; punches (perforators)—round, oblong, square, rectangular, hexagonal, triangular, etc.; quills for small punches; die buttons; die plates (retainers); punch plates (retainer plates); backing plates; round drawing dies; round drawing punches; stripper guide bushings; dowel pins; stop pins (gauge pins); trigger stops; retractable, elastic stops; riser pins; ejector pins (knockout pins); pressure pins; shanks; springs (steel, rubber, urethane); spring plungers; spring retainers; oil film breakers; pilot pins; starting stops (finger stops);

roller stock pushers (spring rollers); spring guides and guards; scrap cutter units; die-handling hooks (transportation hooks); dry seal pipe plugs; screws, bolts, nuts, washers, etc.

Company Standards

Many industrial companies (especially the larger ones) have their own house standards for several phases of die design.

CHAPTER FOUR

Presses and Their Care

The large variety of presses employed in press shops may be classified roughly according to:

a. Source of power: manual (flywheel presses), mechanical (nongeared and gear types), hydraulic (water or oil pressure).

b. Method of actuating the slides: crankshaft, eccentric gear, friction screw, knuckle joint, toggles.

c. Number of slides: single action, double action, triple action.

d. Press frame: gap ("C"), straight frame side; and by position: vertical, inclinable, inclined, horizontal.

e. Clutches: mechanical (positive or pin dog), friction, eddy current.

f. Brakes: mechanical, disc or drum type, dynamic.

Each of these types of presses has characteristics that make it more suitable for certain operations than for others, as will be seen below.

Figures 4-1 to 4-5 illustrate a few models of presses built by some of the leading U.S. manufacturers. In addition, presses and press-brakes are shown in several illustrations, which appear in some of the subsequent chapters.

Press Selection

To select the proper press for a given job, several factors must be taken into consideration in order to achieve best results from the standpoint of efficiency, economy, capacity, etc. The chief factors may be listed as follows:

type of operation

stamping forces required by the job

work capacity of the press

dimensions of the press as required by the stamping tool and accessories (shut height, slide and bed area, bolster plate openings, throat distances, shank hole in the ram, etc.)

work requirements

operating speed (i.e., strokes per minute) and production rate

stroke length

feeding and ejecting devices

blank-holding devices

Fig. 4-1. 250-ton stamping press. (*Courtesy of Clearing, a US Industries Company*)

Operations

The type of operation to be performed and its characteristics pretty much determine the type of press to be selected. In fact, the basic types of press operations require more or less definite press features.

Cutting operations (blanking, punching, trimming, notching, etc.) of thin and medium gage stock and average (small to medium) size stampings need short-stroke mechanical presses with flywheels (nongeared drive). For heavier stock and/or larger stampings, back-geared mechanical presses (or eccentric shaft presses) are preferred.

Note. First operation tools require higher speeds (stokes per minute) than second operation dies of the same general characteristics.

Fig. 4-2. 1000-ton stamping press attended by a team of press operators. (*Courtesy of Clearing, a US Industries Company*)

Bending and forming operations require presses with rather long strokes; both mechanical and hydraulic presses are used (often press-brakes and, for very light duty, flywheel presses).

Drawing is usually done with mechanical (nongeared or back-geared) or hydraulic presses (the latter are preferred, especially for very deep draws); slower spm and longer strokes are usually needed than for cutting. In addition, for very deep draws and severe operations, special double-action (sometimes even triple-action) presses are used to provide the necessary blank-holding pressure, for some supplementary forming or cutting operations, or simply for ejecting the finished shells.

Compression Operations, which demand heavy plastic flow of metal (coining, swaging, sizing, embossing, cold-extruding, etc.), are best done with hydraulic presses that give constant, steady, uniform pressure at any point of the ramstroke; however, quite often rugged mechanical presses (knuckle-joint, toggle, friction presses, etc.) are used.

Desirable Features of Presses

When choosing between existing presses, for a particular job or when buying a new press or when an obsolete press is being considered for modernization, then the following characteristics are of importance:

Fig. 4-3. 60-ton fully equipped OBI press. (*Courtesy of The V & O Press Company, Inc.*)

1. Sufficient strength and rigidity (especially in the frame) to prevent some member from bending too much or bending under the anticipated working stresses. Adequate weight and appropriate design to minimize any danger from vibrations due to impact stresses. In case of weak machines, even if the press itself does not suffer during operations, the dies are endangered and the workpieces are usually imperfect. In addition, tool life is drastically reduced. All this because a weak press deflects too much under stress and cannot properly guide the ram and thus the die cannot work with the necessary accuracy.

2. Safe and possibly noiseless clutch and powerful brake.

3. Large punch-holder ram with extra-long guiding surfaces with accurate fit between ram and guides.

4. Centralized lubrication; better if fully automatic.

5. Equipped with nonrepeat device.

Fig. 4-4. 150-ton high-speed Hydraulic Performer®. (*Courtesy of Pacific Press & Shear Company*)

6. Large flywheel for increased work capacity (of course, always in correct relation to the general design and construction of the press).

7. Ram of symmetrical design which ensures uniform distribution of weight and stresses.

8. Inclinable body.

9. Knockout crossbars.

Fig. 4-5. 300-ton hydraulic punching and blanking press. (*Courtesy of Pacific Press & Shear Company*)

10. Ease of handling, both in operation (controls) and in adjustments. In case of foot-pedal control, it is good practice not to have to make more than a 2-inch pedal travel and pedal actuation should not require more than 20 pounds in case of female operators and 30 pounds in case of male operators.

11. Ample margin of adjustment in the position of the ram and any other member which has a regulable feature such as table height.

12. Accuracy of working, especially concerning the ram and the press table (precise perpendicularity is an essential condition).

13. Long working life of the press assured by ample proportions of all parts subjected to wear.

14. Easy and low-cost maintenance.

15. Best suited material for each part and overall first-class craftsmanship in the building of the press.

16. The driving electrical motor should be of the special high-torque, high-slip type. These motors have high starting and accelerating torques and sufficient slip under load to allow the flywheel to deliver its kinetic energy at the right moment.

Inclinability of Presses

An inclinable design offers several advantages that are not always appreciated and taken advantage of. (It is a common occurrence in metal stamping shops to see inclinable presses used permanently in the upright position.) The main benefits obtained by inclination of presses are:

Possibility for automatic feeding and automatic unloading of stampings and slugs.

Increased productivity and decreased accident hazards.

Oil and other lubricants drip away from the operator, instead of toward him, if the press is flooded.

Special Features of Presses

Since presses can be equipped with many different extra *attachments* or presses may be built for special purposes with corresponding characteristics, it is always good practice to consider present *and* future needs in this respect.

The usual (almost standard) attachments are slide air counterbalance, die cushions (for drawing and forming), tie rods (for medium size and large presses), variable speed drives, stroke counters (preferably electrical), power press-ram adjustment, and every kind of feeding and ejecting devices, scrap cutters, etc.

Special purpose presses may be of the following features or types: deep throat, adjustable bed height, transfer presses, horning presses, horizontal presses, multiple slide presses, twin feed tables, hydraulic double end presses, etc.

Evaluating Stamping Forces

Most people connected with metal stamping production often use the expression "stamping pressure." This is a misnomer. In fact, when you speak about cutting pressure (or bending pressure or drawing pressure, etc.) you do not mean "tons per square inch" or "pounds per square inch," which are units of pressure. Instead, You are talking about "tons" and/or "pounds," which are units of *force*. So, the correct expressions must be: cutting force, bending force, drawing force, etc.

The case is different when we speak about operations involving massive deformation (coining, cold or hot forging, cold or hot heading, etc.) There, the expression of "specific stamping pressure" (for a given material) is justified partly because the required amount of "pounds per square inch" (a unit of pressure) is the basis for the calculation of the total force needed. (Total force = specific pressure × square inches of surface to be deformed.)

For the determination of the stamping (cutting, forming, drawing, compressing, etc.) forces required for performing a given job, the universally known and readily available formulas, nomograms, tables, etc., are used, as may be found in handbooks, manufacturers' engineering catalogs, etc. However, it must be clearly stated that all of these give only approximate values. Therefore, in order to stay on the safe side, it is recommended that an ample safety margin be added to the previously calculated values and thus a press selected with ample capacity. For example, if the computed total stamping force is about 100 tons, choose a press of at least 150, and better if 200, tons nominal capacity. The larger the press, the less the strain on it, consequently, the less the noise and vibration caused by the operation. In addition, tool life is increased considerably in this way, because of less deformation.

During the force requirement determination also take in account every attachment that must be actuated simultaneously with the performing of the stamping operation (blank holders, ejector-pressure pads, strippers, etc.); as well as the force required to push slugs or blanks through straight-walled openings, as sometimes happens.

Since the actual capacity of a crankshaft press, which is usually used in press shops, is greatly dependent on the distance above the bottom of the stroke where the force starts being applied, this factor must be considered (heavy blanking, forming, drawing, etc.).

When choosing a press of suitable capacity, it is not enough to check the force requirements; it is also necessary to take in account the energy (quantity of work) done during the stamping operation.

When selecting a press for a new job, it is good practice, whenever possible and feasible, to compare the new job with some previous job having similar characteristics. In this way much trouble is often avoided.

Accuracy Requirements

The accuracy required by a given job has the following influence on the press selection: If the tool is built with rather close tolerances (especially with respect to punch-to-die clearance), it is necessary to choose a press that should have the least deflection under load. This condition is achieved by either selecting an open gap frame press with an ample capacity margin (over 100%) or a straight side press, which is intrinsically more rigid.

Speed of Presses

With reference to presses, the term "speed" is often used incorrectly. Speed means basically the distance covered by a moving object during a given unit of time. However, it is customary to call the "speed" of a press the number of double strokes (one travel down + one travel up) per minute it makes while running continuously. In the case of crankshaft presses, this number is independent of the stroke length, even if this is adjustable. In the cases of screw presses (both friction presses and screw fly presses) and hydraulic presses, the "speed" depends on the length of the stroke; the longer the stroke, the lower the number of stokes per minute and vice versa. Consequently, for friction presses and hydraulic presses the stroke length should be the shortest possible, compatible with the requirements of the corresponding press operations.

The author will use instead of "press speed," simply the initials "spm" (strokes per minute).

The optimum spm of a press in a given case depends on several factors: size of press, type of press drive, size and weight of the stampings, physical properties of stock, kind of operation, stroke length, stock thickness, center distances, and feeding and ejecting methods.

As a basic principle, the larger the press and the larger the stamping, usually the slower the press. The following data are good average values:

Press size

heavy presses (over 100 ton)	3–35 spm
middle size presses (30–100 ton)	25–100 spm
small presses (under 30 ton)	100–350 spm or more

Note: There are reports of presses operating up to 1000 spm and even more which give excellent performance.

First operation cutting is usually faster than second operation cutting and forming. Heavy stock (and/or stock with higher shearing strength) requires lower spm than thin stock

over 0.250-inch stock thickness	25–75 spm
0.080–0.250-inch stock thickness	50–200 spm
under 0.080-inch stock thickness	150–300 spm

Shorter center distances in the strip allow faster feeds and, therefore, higher spm values and vice versa.

For *bending* operations with simple "V" tools the press spm should bear a certain relation to stock thickness:

thickness	0.010	0.020	0.030	0.040	0.060	0.080	inch
spm	80	70	60	50	40	30	

Drawing must be effected slowly in order to allow time for proper plastic flow of materials. The following maximum *lineal velocity* (speed of the ram when the drawing begins) values are recommended for single action drawing:

brass	200 fpm
aluminum	175 fpm
copper	150 fpm
zinc	150 fpm
steel (drawing quality)	55–80 fpm
hard aluminum	40 fpm
stainless steel	35 fpm

Variable Speed Drives

For every job there is an ideal spm that is the best from every standpoint (production rate, tool life, press maintenance, accuracy of stampings, etc.), but it is practically never possible to find a press in the shop with just that spm value, although recently in several shops very successful experiments have been carried out with variable speed drives. These allow finding the most satisfactory spm to meet a given combination of job characteristics, such as type of operation, differences in materials and stock thicknesses, feed lengths, ejection, etc.

In many cases the rpm values of the drive can be modified within very ample limits and often without the necessity of stopping the press. This feature is particularly valuable in those cases where several presses must be synchronized in order to get a uniform, constant flow of automatic production.

Variable speed drives have, today, many enthusiastic advocates; however, a word of caution must be mentioned: speed cannot be changed indiscriminately. The capacity of a press varies with the square of the spm. For example, if a press with an average 30 spm is reduced by 50% to 15 spm, its rated capacity is reduced by a ratio of four to one. On the other hand, when the spm is increased by 50% to 45 spm, the rated capacity is increased by a ratio of 2.25 to one.

When changing the spm of a press, all other controlling factors must be taken into consideration, such as press strength (in case of an spm increase), stroke length, tooling, material, feeding equipment, etc.

Dual-Speed Press Drives

Another method for improving press performance has been developed by several press manufacturers which increases the spm value considerably. Essentially, this method consists of providing "differential" velocity, i.e., the ram moves quickly when it does not perform any work (during the approach and return movements) and it moves slowly during the actual performance of the stamping operation (analogous to the "quick-return" motion of shapers and similar machine tools).

The practical solution of the problem in question may be effected in several ways (usually mechanically), but the working principle is always the same.

Production Rate

Intimately related to the operating speed is the detail of the operational cycle that should receive the greatest attention, i.e., the production rate of the actual productive phase (the transformation of the raw material into an interchangeable component). For best production efficiency we must produce in a given lapse of time (hour, day, week, or shift) the maximum quantity of finished stampings of satisfactory quality.

It is an established fact that from the standpoint of production efficiency, press operations are among the top metalworking processes. Metal stampings are produced very quickly, usually in fractions of minutes (sometimes, fractions of a second). However, this does not mean that we should ignore the possibility of improving production efficiency by increasing the production rate to an optimum.

The possibilities of maximum production rates for a given stamping depend on several interdependent factors such as: type of operation and characteristics of the tools; size of the stampings; characteristics of the stock, of the available press, and of the available auxiliary equipment (chiefly feeding and ejecting); and other secondary factors.

The maximum attainable production rate is usually determined by the speed of feeding, which often may represent a real bottleneck. This in its turn is limited in manual feeding by the operator's dexterity and by fatigue and in semiautomatic and fully automatic feeding by the conditions of forces, weight, and inertia involved, which can provoke breakdowns and work interruptions, if they become excessive.

As a first approximation, the feeding method has the following effect: semiautomatic and fully automatically fed, first operation dies may catch every press stroke, while the press is running continuously; manually fed first operation dies have a production rate of only about 50% of the nominal spm of the press. In the case of second operation dies the actually utilized number of strokes is about only 15–30% of that of the corresponding first operation dies.

As already stated, in second operation dies (usually single stroke operations) the absolute value of the press nominal spm is practically immaterial when compared with the lost or secondary time intervals needed for the manipulation of the stampings and the press before, during, and after the operation. The production rate may be much more effectively increased by reducing the lost-time intervals. The tendency should always be toward transformation of the single-stroke technique into a continuously running one. This fact does much more to improve production efficiency than an increase in press spm. The total elimination of the stresses of continually repeated accelerations and brakings (continuous running instead of single-stroke technique) also results in the additional advantages of longer press life and reduced press maintenance.

Press Performance

If one can choose among several presses for a given job, then, as a first approximation, the answer would be to take the press with sufficient capacity (ample margin of safety against overloads), with sufficient table area and shut height, and with the highest spm value.

This is satisfactory as long as it is a case of a die equipped with automatic (or, at least, semiautomatic) feeding devices, so that each ram stroke (each flywheel turn) can be made to work effectively; otherwise, there may be trouble. In the case of single-stroke technique, such as hand-fed first operation and/or second operation tools, the increase of press spm has little or no influence on the production efficiency and sometimes it even has negative consequences.

Take two comparatively small presses, one having 120 spm and the other, twice as much, i.e., 240 spm. The actual time of a complete ram stroke is 60 ÷ 120 = 0.5 sec in the first case and 60 ÷ 240 = 0.25 sec in the second case. A superficial judgment would consider the production rate ratio to be one to two. However, it is actually not so because we must consider the time needed for a complete operational cycle, i.e., including the feeding and removal of the workpieces. Taking 3 sec for loading and unloading as an average for a second operation job, the total operational cycle would be 3.5 sec with the slower press and 3.25 sec with the faster one. So the actual ratio is only 1 to 1.08, i.e., there is practically no increase in the production rate. In some cases the higher spm may even be harmful, as will be demonstrated in Chapter 6; it could mean also excessive, unnecessary wear of the equipment. The improvements must be sought in other factors and effort must be exerted to reduce the secondary and lost-time itervals. For more information about this matter, see later in this chapter and also Chapters 10 and 11.

Stroke Length

As was previously mentioned, the required length of ram stroke depends on the type of operation to be performed as well as certain tool dimensions. As a matter of principle, the press should always have the shortest stroke compatible with the work requirements. This is contrary to current practice where presses usually have exaggeratedly long strokes. This has several drawbacks, the chief one being that long strokes usually, but not necessarily, result in lower spm and thus slower production cycles. Besides C-frame presses will have considerable deflections and consequent tool misalignments.

A cutting die should be operated with the shortest possible stroke. This is determined by adding an allowance of about 0.125 inch to the strip feeding canal height (see Figure 4-6). Usually full stroke length values are 0.50 and 0.75 inch. On the other hand, the stroke must be considerably longer if some feeding, ejecting, or other attachments must be actuated. In fact, the rate of such automatic operations is usually governed by the feeding time and, with longer strokes, a greater percentage of the operational cycle is available for feeding. The presence of pilots which require somewhat longer press strokes than dies without pilots must also be taken into account.

In second operation jobs, long strokes are often needed in order to permit easy location and removal of the workpieces.

For drawing, forming, or bending, the ram stroke must be quite long, in any case more than sufficient to complete the operation. For example, in drawing of flangeless shells, it is recommended that a press whose stroke is or can be adjusted to 2.5 to 3 times the depth of the deepest shell be selected.

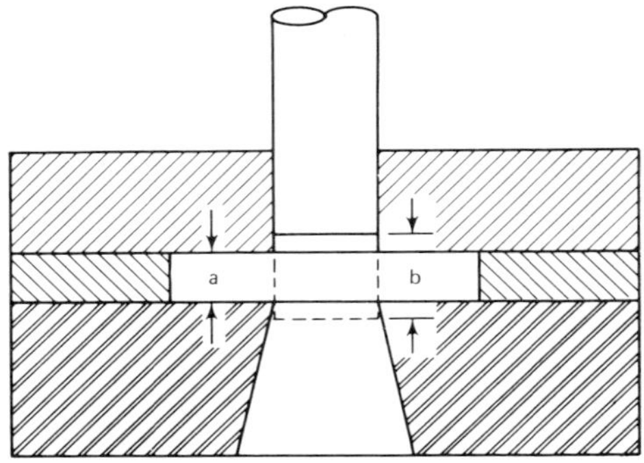

Fig. 4-6. Low feeding canal height for cutting operations.

Care and Maintenance of Punch Presses

Obviously, the production efficiency of a metal stamping press shop depends to a great extent on the working conditions of its equipment, chiefly presses. Another very important detail, maintaining production and delivery schedules, also depends on the efficient operating conditions of the production equipment.

Keeping presses in first class working condition calls for more inspection and maintenance than for many other types of production machines because presses are called upon to take greater punishment and are subjected to very high stresses, due to their operational principle, i.e., rigid ram guiding with the tool presenting a sudden, very high resistance. The sudden starting and stopping of heavy masses of the machine at the beginning and at the end of each stroke together with severe impacts and vibrations highly stress each part of the press. Play in the guides and bearings worsens this situation. Therefore, presses need more care than other production machines or machine tools, other conditions being equal. Furthermore, press tools are worth hundreds or even thousands of times more than tools used in ordinary machine tools, and a press in bad condition shortens considerably the useful life of the dies.

Press breakdowns may be classified according to their causes into two groups: those due to mechanical failures and those due to human causes. The first one (natural wear of parts, faulty design, careless craftsmanship, defective materials, accidents) belongs entirely to the province of the maintenance department.

The second group (lack of care; inadequate operational technique; overloads of every kind, two parts in a die simultaneously; slugs under the operating parts of a die; die member misalignment; or faulty lubrication) can only be avoided by proper care exerted by operators, die setters, foremen, supervisors, and maintenance personnel.

It pays to take care of a press. With very little attention, effort, time, and expense, the press may be kept in excellent condition until it becomes obsolete, thus conserving valuable equipment, avoiding breakdowns, accidents, loss or interruption of production, and large expenditures for repairs.

Press Maintenance in General

It is a sad but established fact that only a relatively small percentage of companies consider the task of machine (plant, equipment) maintenance with the same concern as the production efficiency. In the majority of plants, the plant engineer deals only with emergencies, i.e., breakdowns or actual production interruptions. Many companies believe that as long as their maintenance department handles breakdowns promptly and efficiently, holding down production interruptions, they are operating satisfactorily. In quite a few cases, because maintenance is not a direct productivity activity and its contribution to overall plant efficiency (and to profits!) cannot be evaluated in monetary terms, easily and immediately, it is considered as a necessary evil that eats up profits. Consequently, the maintenance department may be assigned only second-rate personnel and second-rate equipment.

In the opinion of progressive minded people, this kind of policy is a grave and unpardonable error, because the efficient operation of a plant does depend to a great extent on a correctly planned, organized, and well-functioning maintenance department. Only in that way can the production equipment be kept in excellent shape.

Even if one considers the matter from a purely economic or financial standpoint, it is found that downtime due to breakdown usually costs much more than direct maintenance. The economic consequences of an unscheduled production interruption are always unpleasant and may be disastrous: schedules are disrupted, machines and workers are idle, production is decreased, customers are lost.

Preventive Maintenance vs Corrective Maintenance

Analogous to modern progressive medical practice, modern maintenance policy of plant engineering puts the greatest emphasis on *prevention* of breakdowns instead of repairs. Extensive studies of actual plant operations have demonstrated that the overwhelming majority of mechanical failures of equipment do not happen suddenly, but are cumulative in nature; failures come about progressively. Almost always, some small, seemingly unimportant defect presages the possibility and probability of serious mishaps. (Early symptoms are an invaluable help for preventing major failures.) Surveys to determine the causes of actual breakdowns have established that in over 75% of the cases the cause for a complete breakdown was to be found in faulty lubrication; in the breakage or failure of a packing, a brake or clutch lining, a switch, or a relay; in a loose screw, nut, or bolt; or in some similar trifling detail. Therefore, if a serious preventive or "anticipated" maintenance program is conscientiously carried out, such small defects can be detected in time and eliminated promptly, and, thus, breakdowns can be avoided or, at least, minimized with practically no interruptions of production.

It should be pointed out that preventive maintenance has been proven in actual practice to be inexpensive and economical. There is no need for large investments in capital equipment. Even the smallest shops can easily afford the installation of an efficient preventive maintenance organization.

The goals, purposes, and objectives of a well-planned and correctly carried out preventive maintenance program, may be summarized as follows:

1. To maintain the plant in efficient operational conditions for uninterrupted productivity and, if possible, even improve its production efficiency. (Any unscheduled interruption decreases production efficiency as will equipment not in top operating condition.)

2. To avoid or, at least, to minimize breakdowns to an absolute minimum by discovering in time any conditions that may develop into equipment failure. Production efficiency is held at its maximum by replacing worn, spoiled, or damaged equipment components before they cause serious consequences.

Unwanted shutdowns cause loss of production (decrease of output), cost increases (downtime, idle operator time, spoilage, rework), impaired employee morale; decreased overall production efficiency; repair costs that are higher than preventive maintenance costs, because the latter catches the defects while they are still small; and, if the breakdown is prolonged, then lay-offs become necessary and there is the risk of losing well-trained, valuable employees.

3. To improve working conditions. Operators have the right to expect that equipment will be kept in such condition as will permit satisfactory production, both in quality and quantity. (Apart from the economic factor, operators want to have satisfaction in their work.) In addition, in a well-maintained plant, supervision is apt to be much easier.

4. To ensure safety for the operators. It has been found repeatedly that severe accidents still occur due to mechanical failures. The majority of these can be traced to the lack of proper maintenance of the press itself. This demonstrates the necessity for increased vigilance and frequent periodic inspections of those parts that are most highly stressed mechanically, in order to ensure that defects are discovered and taken care of before they cause an accident.

5. To prolong the useful life of production equipment (lower depreciation of plant assets). Very often it is impossible to distinguish between the consequences of natural wear and those of carelessness. The premature need for overhauling of a press—either because of major irregularities in its working or because of lack of accuracy in its production for whatever reason—means unnecessary expense, loss of production, and decreased productivity.

6. To increase the efficiency of the maintenance department personnel. With a systematic preventive maintenance program, one press after another is inspected, serviced and repaired. Thus a small repair crew can take care of a comparatively large number of presses. Also, programmed preventive maintenance helps to avoid the need for any excessive overall maintenance activity.

A well-organized and correctly cared for plant makes a better impression on visitors who know something about maintenance and management.

On the other hand, although in most cases preventive maintenance is a great improvement over "shutdown" maintenance, the former should never become a goal in itself. Time is required for inspections, for a well-kept record system, and the analyzing these records. All this costs money. There are some special cases where the risks of loss of unscheduled production interruptions are low, so it may cost less to use only corrective maintenance or a combination of preventive and corrective maintenance. All depends on the comparative overall costs.

By the way, when you start organizing a preventive maintenance program from scratch, do not expect immediate results. It usually takes at least 6–8 months until the program begins to show some tangible results.

Preventive Press Maintenance Check Lists

Preventive maintenance has three distinct—yet intimately intermingled—aspects: *inspection* (do not wait for a failure to happen; detect it beforehand), *servicing* (it pays to look after a machine), and *repair* (if inspection finds something wrong, repair it immediately). All these activities, of course, must be scheduled in advance according to proper plans.

The actual press-maintenance program msut be elaborated for each press shop individually, taking into account the characteristics of the presses in question and the particular conditions and requirements of the shop, according to the type of usage, quality of personnel, etc. There are a few basic concepts that have general applicability and must be given due consideration.

Only through rational planning can an efficient, systematic preventive maintenance program be effected. The very core of it is the *check list* of the fixed periodic inspection and servicing activities. Practically all preventive maintenance revolves around the check list. Such a list for regular, routine inspection and servicing must be complete but not overcrowded; no important, critical detail should be overlooked, but, on the other hand, the inspector should be able to cover a comparatively large number of presses and other operating equipment in a given time. The check list must indicate what to do, when to do it, how to do it, and who should do it. It must also include a program for the replacement of worn parts, overhauling, etc.

The preparation of such check lists is not an easy job. The best way to proceed is to consider the directions and recommendations of the respective manufacturers; but equally important and useful is consultation with the operators, diesetters, foremen, and maintenance personnel who are familiar with those parts of the machines liable to give trouble. In addition, the supervisors of the department where the machine is installed should be consulted. Written reports of past experiences are also extremely useful. Sometimes, it may be advisable to call in special consultants (industrial engineers) for this job.

Periodic preventive maintenance tasks, once established, usually are performed as a matter of routine, according to the prepared lists. However, it is recommended that the routine be checked from time to time in order to see whether or not the maintenance is enough, too much (which does not do any harm, except that it costs money), or too little, and, if necessary, to modify the check list accordingly.

As previously stated, each shop, especially with its different machinery (type, size, working conditions, and, above all, age), requires different preventive maintenance procedures. Nevertheless, a brief outline, which could be taken as a skeleton for the establishment of a preventive press maintenance program, follows.

The outline is divided into five phases: daily, weekly, monthly, quarterly, and yearly. Whether a press is in use constantly one shift per day or when used only occasionally (i.e., with very low machine-hour utilization factors), it is not the calendar that determines each phase, but the number of operational hours (machine hours). So, the five phases may be defined in this fashion: every 8 hours, every 40 hours, every 160 hours, every 500 hours and every 2000 hours of operational time, respectively. In case of doubt or insufficient information about previous performances, assume higher utilization factors than what the records indicate. An error possibly will mean too much maintenance activity, but that is preferable to too little.

Daily Maintenance consists of cleaning, lubricating, and visually inspecting the machine at the start of a shift. See Chart 4-1.

Chart 4-1: Preventive Maintenance Daily Treatment (every 8 hours of operational time)

1. Cleaning

The sequence is very important. Start with a clean rag on the sliding surfaces (ram, gibs, etc.); afterward, clean the table and finish with the press body (uprights, legs, etc.).

2. Lubrication

Grease the leader pins of the die set. Oil the sliding members of the tool. Remove any excess oil or grease.

In case of presses *not* equipped with a centralized automatic lubrication system, oil every bearing, guide, gib, etc., of the press. Proceed systematically, following exactly the established procedure of the check list, in order not to miss anything. Never employ the same brush for lubrication of the stock that is used for lubricating the press or the tool.

3. Visual Inspection

The following details musts be checked carefully and, if necessary, adjusted properly:

a. Clutch operation
b. Brake operation
c. Ram clearance
d. Bearings
e. Belt tension
f. Treadle spring tension
g. Air connections (for leaks)
h. Counting devices
i. Electrical, hydraulic, and pneumatic controls
j. Safety devices (guards, pullbacks, treadle guards, flywheel guards)
k. Illumination of the work area

Weekly Maintenance. Once a week every press must be checked thoroughly and, if found faulty, repaired. Chart 4-2 lists the items to be checked. After completion of the necessary repairs, controls, and adjustments, this chart is filed in the corresponding press file (held in the maintenance department office).

Monthly Maintenance is carried out according to Chart 4-3.

Quarterly Maintenance is detailed in Chart 4-4.

Yearly Maintenance. First of all the gear housing (cases) and all lubricant reservoirs should be drained, cleaned, and refilled with the proper type and amount of lubricant. In addition, the press should receive a general overhaul consisting of stripping all parts and cleaning and repairing them as necessary. It is good policy to also include in the yearly preventive maintenance program a magnaflux (or similar) test of the crankshaft and of the pitman screw.

Chart 4-2: Preventive Maintenance

Weekly Press Inspection Report (every 40 hours of operational time)

Machine #________	Date ________
Flywheel ________	Slides (alignment, clearance, etc.) ________
Clutch ________	Pitman and locking nut ________
Clutch wheel ________	Gibs ________
Clutch plate ________	Screw ________
Clutch linings ________	Counterbalance ________
Pull-out dog ________	Air cushion ________
Springs ________	Counters ________
Pins ________	Frame (fractures) ________
Dogs ________	Tie rods ________
Treadle spring tension ________	Excess lubrication ________
Brake:	Insufficient lubrication ________
lining ________	Lubrication lines ________
band ________	Mounting ________
springs ________	Air valves and air connections ________
stud bolts ________	Safety devices ________
adjustment ________	Guards and covers ________
Bearings ________	Electrical:
Gears ________	motor ________
Crankshaft ________	controls ________
Belts and pulleys ________	hookup ________
Air leaks ________	switches ________
Packings ________	buttons ________
Loose bolts ________	pedal ________
Levelness ________	

Note: Any information pertaining to what should be done to the press and/or what has been done to the press during inspection shall be reported on the reverse side of this form. Give a complete description.

It is sound practice to combine the overhauling with some rebuilding and/or modernization. Such overhauling should follow the methods employed by conscientious used-machinery dealers and by railroads. It is recommended that any overhauling and modernization be done by some outside agency, i.e., qualified press rebuilders or, even better, by the original manufacturer of the press.

Modernization is indicated mostly for large presses. It may include the improvement of several press details, chiefly changing old-fashioned pin-type (positive-type) clutches to up-to-date air friction clutches; changing bronze bearings to roller bearings; installing special drives (for fast approach speeds, slow draw and quick return); application of automatic (or, at least, centralized) lubricating systems (instead of the existing manual ones), etc. In addition several devices may be added to increase production efficiency, such as feeders, knockouts, air cushions, safety overload devices, counterbalances, motor-driven slide adjustment, etc. For all these items consult the press manufacturer, who is in the best position to give the most useful advice.

Chart 4-3: Preventive Maintenance

Monthly Press Inspection Report (every 160 hours of operational time)

Machine #____________________ Date ____________________

Lubrication. Check all bearing points for lubrication.

Drive assembly and clutch. Check for leaky packings. Check for worn or improperly adjusted belts.

Ram. Check gibs for proper clearance and alignment. Check elevating limit switches for proper operation.

Counterbalance cylinders. Check for air leaks. Change packings, if necessary.

Air cushions. Check for air leaks in packings and pipings.

Electrical motor. General, overall check-up.

Note. Any information pertaining to what should be done to the press and/or what has been done to the press during inspection shall be reported on the reverse side of this form. Give a complete description.

Scheduling

Except for the daily maintenance activity, all others must be properly scheduled on a rotating basis and so staggered that long, simultaneous production interruptions are avoided. Such a staggered schedule also helps to keep the maintenance department personnel roster comparatively low.

The master schedule is prepared in this way: Past performance files show the average time utilization per week of each press so that the 40-, 160-, 500-, and 2000-hour jobs can be easily established (calendarwise) for each press. Then a master chart is prepared on which the single preventive maintenance jobs are posted in their proper time location, at the right intervals.

Of course, it is practically never possible to follow the master schedule exactly. There will always be some unscheduled, unexpected occurrence that will necessitate changing (moving up or postponing) some inspection job. This means that some reshuffling or rescheduling has to be done. Consequently, the master schedule must never be considered as fixed and rigid; it must be flexible, and changeable within reasonable limits. Furthermore, the master schedule should be revised and remade about every month or so.

It is recommended scheduling the long interval inspections (500 and 2000 hours) so that they be performed during off-season periods. Thus the regular production does not suffer. The basic idea is to conduct most of the preventive maintenance activity during a time when the press would normally be idle anyway. In this way costly shutdown time is avoided.

In any case, there should exist a very close collaboration between Production and Maintenance. Every scheduled and programmed shutdown of presses for maintenance must be previously approved by Production, in order not to disrupt production schedules. On the other hand, Production Plan Schedules must foresee and allot time for preventive maintenance activities, possibly within the framework of the Maintenance Master Schedule.

Chart 4-4: Preventive Maintenance

Quarterly Press Inspection Report (every 500 hours of operational time)

Machine #__________________ Date __________________

1. Remove flywheel.
2. Take apart clutch—clean, inspect, and adjust.
3. Take apart brake—clean, inspect, and adjust.
4. Remove pitman, ram, gibs—clean, inspect, and adjust.
5. Check main bearings for wear.
6. Check frame for cracks.
7. Drain lubricant from reservoirs, clean out, and refill.

Note. Any information pertaining to what should be done to the press and/or what has been done to the press during inspection shall be reported on the reverse side of this form. Give a complete description.

Obviously, complete, detailed service records must be kept of every maintenance job, according to Charts 4-2, 4-3, and 4-4.

Organization of the Preventive Maintenance Department

About the question of who should perform the actual preventive maintenance work, there are no hard and fast rules, and there is no general agreement; opinions and experiences are rather diversified. However, the following ideas are quite common and are carried out in several well-organized, well-managed shops.

During daily maintenance, cleaning and lubricating should be performed by floormen and visual inspection by foremen or diesetters. All this according to a proper plan: every serviceman in charge of a section of presses. Of course, this means a staggered timetable; the servicemen must start earlier (one hour or more) than the press operators.

Usually in small shops there is a different routine. The complete daily maintenance (cleaning, lubricating, and visual inspection) is performed by the operator of the press, either at the end (rarely) or at the beginning of the shift (the more usual case). This practice makes much sense. A good press operator, after a few years of experience, becomes quite an expert in his or her job and knows the operation of the machine thoroughly. It is good policy to take advantage of this fact, asking for his or her cooperation, especially for equipment maintenance.

On the other hand, this practice may cause problems in shops with two or more shifts. Some operators may be tempted to assume that the people of the next shift will take proper care of the daily servicing and, therefore, such operators may become slack in carrying out their own duty. In addition, they may assume that personnel of the other shift will report to the foreman any and all irregularities and problems with the machine and, consequently, this important reporting is not done.

The jobs of the 40-, 160-, 500-, and 2000-hour servicing take more time and require more skills and specific knowledge, training, experience, and practice. Therefore, they must be performed by trained mechanics who are familiar with press construction or by qualified, properly trained, first-rate personnel. Sometimes, the diesetter's assistance is also taken advantage of. For major overhauling, as mentioned before, it is best practice to use outside, specialized help.

There are shops where the daily maintenance's visual inspection is not carried out by the operator (or diesetter or foreman) and not at a set time (e.g., at the start of the shift). In these cases there are so-called "patrolling" inspectors (skilled mechanics) who inspect all the presses assigned to them, following a fixed routine, performing the checking of one machine after the other. (Such inspections are rather short: the average time needed is only 10–15 minutes per press.) These people soon become familiar with the equipment to which they are assigned, a familiarity substantially reinforced by the casual conversations held with the corresponding press operators, during which they often get very valuable first-hand information about problems. These maintenance personnel are then able to detect sources of trouble at an early stage and to correct them on the spot. Major problems are reported, and corresponding repair orders are requested.

In the case of rather large plants, with a well-staffed maintenance department, it is recommended that the people assigned to preventive maintenance be separate from those who carry out emergency mantenance in case of accidents (repair crews). Such specialization is very valuable and useful since their respective jobs are rather different in nature and skill required. Suggested proportions of preventive maintenance to emergency maintenance personnel are two to one, or three to one.

As mentioned before, it is very important and useful that correct training be given to the maintenance personnel. The old-fashioned maintenance mechanic with his toolbox containing a hammer, a chisel, a few screwdrivers, a pair of pliers, or a few wrenches is no longer seen. The maintenance mechanic of today must be an engineering technician, well trained not only in machinery as a whole, but also in electricity, electronics, hydraulics, fluidics, etc., in order to be able to deal with the sophisticated controls of today's presses and their accessories. It is good policy to send the maintenance people (or at least their foreman) to the press manufacturer's plant, in order to receive training in the care and maintenance of their presses.

Service and maintenance instruction manuals for every piece of equipment should be kept on hand in the maintenance office and should not be filed away in some obscure general office files. These manuals should be required reading for all maintenance personnel. Machine and equipment manufacturers are always anxious to distribute their instruction manuals among people who need them (especially to users' employees). These manuals are usually very well prepared and serve their purpose excellently.

Obviously, in order to be able to operate efficiently, the maintenance people should be provided with the best available working equipment, such as tools, machines, material handling means, and spare parts.

In order to keep the shutdown times caused by repairs and/or substitution of worn or damaged components low, thus improving the efficiency of the maintenance department operation, it is necessary to keep a well-stocked spare parts store. The determination of such spare parts inventory is a very difficult job, because of the innumerable unknown and uncontrollable factors involved. The only real help comes from past experience, together with suggestions and recommendations of the press manufacturer.

The press maintenance department should be located strategically, i.e., as near as possible to the center of the area to be serviced.

Press Accuracy Inspection

To preventive press maintenance belongs also the detail of precision inspection. Regular routine inspection for accuracy should be made at fixed intervals (every couple of months or so). The following conditions must be checked carefully.

1. With a straight ege check the flatness of the bed or table and of the bolster plates. If there is sagging (center depressed because of too many or too heavy impacts), the bed must be remachined and the press subsequently assigned to lighter service.

2. Perfect squareness between ram axis and table surface. To check this, prepare a sweep with an accurately fitting round shank which is secured in the press ram, just like ordinary die shanks. At the free end of the sweep, mount a dial indicator. Lower the press ram carefully until the indicator touches the bed or bolster plate, carefully turn the sweep, and read the indications of the dial indicator. In this way, it may be seen whether or not the press satisfies the chief condition of accuracy, i.e., the variations are between the allowed tolerance limits. Any inaccuracies must, of course, be corrected.

3. Minimum or no vertical play of the ram. Otherwise, at the severance of the slugs or blanks the punch suddenly descends too far causing unnecessary tool wear.

Production Equipment Maintenance

About the other machines and pieces of equipment:

1. Shears and press brakes are treated analogously to presses.

2. Cradles, reels, feeding devices, conveyors, cranes, lift trucks, jeeps, trimmers, welders, rolling and beading machines, machine tools, etc., should be treated according to the respective instructions and recommendations of their manufacturers and according to sound common machine shop pactices.

CHAPTER FIVE

Stamping and Stock

Because this book deals with improving productivity, stampings and stock are dealt with only as far as they affect the easy production flow through the shop.

Stamping Design

The chief characteristics of the stamping (dimensions, shape, stock from which it is made, etc.) determine the required cutting and forming forces. These in turn establish the minimum capacity of the press needed for the operation. In addition, the characteristics of the stamping also directly influence the design of the die, and its size, weight, and other basic features. Consequently, the characteristics of the press to be used are determined also, indirectly, by the stamping.

Stamping design belongs to the province of the component designer; the press-shop people have little or no influence on it, unless, as it should be, their collaboration is requested at an early stage of development, before the final characteristics of the stamping are definitively established. Then, by suitable suggestions, considerable savings of time, labor, and expenses may be realized.

Here, mentioned briefly, are just a few of the highlights of design details; those which affect directly the production efficiency and cost of the stampings:

> Preference should be given to flat (vs bent or formed), small (vs large), simple contours (vs complicated figures), and rounded corners (vs sharp corners) of cutting contours.
>
> Holes in stampings should be produced together with the blank production (progressive or compound vs separate second operations).
>
> Holes in formed stampings should be far from the bending zones.
>
> Springback control; corrective operations should be avoided.
>
> Observe minimum heights of legs in formed stampings.
>
> Compound bendings should be avoided, if possible.
>
> Combined operations reduce costs.
>
> In case of drawn shells, study thoroughly the shape and dimensions in order to avoid difficulties of production.
>
> Tolerances must be as ample as is compatible with part function, in order to avoid costly corrective operations (shaving, restriking, shell sizing, etc.)

Stock Specifications

Logically, it is the component designer who establishes the specifications of the stock to be used for the production of a given stamping. This is done by taking into account the requirements of strength, function, and other criteria. However, it is mandatory that production requirements be considered at an early stage of stock specification.

The recurrent theme of smooth production flow—as may be perceived throughout this whole book—consists of avoiding unscheduled, unwanted, and unexpected production interruptions. Any detail that potentially may provoke such interruptions must be prevented as far as is humanly possible. Therefore, the stock (raw material) must have certain features that are specified clearly when purchased. The chief characteristics are as follows:

1. Adequate quality for the planned press operation. For example, for drawn shells—especially for deep drawn shells—the stock quality must be specially chosen in order to avoid trouble. For plain, flat stampings, the quality requirements are usually simpler.
2. Uniform (high) quality. Variations in material structure may cause trouble in automatic production in case of highly stressed stampings (formed, drawn, etc.)
3. Uniform dimensions. Thickness and width tolerances must be kept as low as reasonably possible, in order to ensure unhindered, smooth feeding of stock and to achieve the required accuracy of the produced stampings.
4. It is decidedly wrong policy to purchase stock that is "just good enough." This is a false economy that often causes much lost time and money. It is far better to spend somewhat more for the raw material, in order to ensure a smooth, uninterrupted production flow.

Stock Preparation

The shape or form in which the stock is presented for production is important.

For first operation cutting tools the stock is utilized in the form of (sheared) strips or coils. The latter are far superior in many respects, including production efficiency. In fact, coils are better suited for full-automatic continuous feeding, thus making possible full utilization of equipment capacity; even when hand feeding coils, there is greater overall production; the smooth burrless edges do not injure the operator's hands; coils are straighter, with less undulations, so there is less chance for punch breakage due to uneven stripping; there is little or no chance of another kind of mechanical short circuit, consisting of half-hole punching at the end of strips and thus punch breakage; presses have a higher utilization factor, because there are more operating strokes per minute; coils are easier to handle and to deliver to the press; they require less storage space, and there is less material waste (both at the side of sheets and at the end of strips some stock is always scrapped).

The economical advantages of coil material vs sheared strips is demonstrated by the following example, taken from actual practice.

Let us take a coil of about 2000-foot length for a die having a center distance of 1.2

inches and mounted in a press of 200 strokes per minute, equipped with an automatic feeding device. If the die is provided with an automatic stop which permits catching every stroke of the press, the whole coil is finished in 100 minutes (24,000 in. ÷ 1.2 in ÷ 200 = 100 minutes).

However, if we worked with sheared strips under the same conditions, we would have the following lost time intervals:

a. lifting the strip	3 sec each
b. introducing it ito the die and registering it against the stop	2 sec each
c. depositing finished strip skeleton	2.5 sec each
d. in case of progressive dies, actuate starting stops 1.25 sec each; let us assume, in our case, two starting stops; so	2.5 sec each
Total each strip	10 sec

Now, the usual length of strips is 6 feet; consequently, to make the same quantity of stampings, we need 2000 ÷ 6 = 333.3 strips. Therefore, working with strips the operator will need 3333 seconds, i.e., 55.5 minutes more than with coil material. This means that there is an increase of 55.5% in operating time and labor.

Even if by such provisions as automatic scrap chopper, longer strips, etc. the lost time is somewhat reduced, there is still a large increase of operating time.

Actually the labor time increase with sheared strips is much more, because straight lengths cannot be fed fully automatically; consequently, not every stroke of the press is utilized for operation.

On the other hand, it must be stated that a change from sheared strips to coils is not always a guarantee of higher productivity. In fact, a rational utilization of coils requires several new pieces of equipment. The material must be brought to the press; coil cradles must be provided; the press must be properly equipped for handling coils (automatic feeding devices); and sometimes even the press tools must be modified, in order to adapt them to the new requirements. Finally, increased floor space requirements must also be satisfied (ample space must be provided around the press for the coil transportation).

Therefore, it is recommended that a thorough study be made in each case in order to establish whether or not it is advantageous to change from sheared strips to coils.

If, for some important reasons, it is impractical or uneconomical to use coil material, then it is recommended that strips be used as long as they are compatible with good stock and sheet utilization factors. Namely, strips are usually hand fed and, in such cases, as practice teaches, each start of a strip takes time, and it also takes time because the operator attains the maximum cadence only after three or more blanks have been stamped (in the case of progressive dies, it costs even more, because the preliminary stages must be superated).

Another detail that must be taken into consideration when shearing strips for blanking, and especially for progressive dies, is this: Punching only half-holes or incomplete holes is very dangerous. The slender cutting points of small piercing punches are pushed away (in the direction of least resistance) and so they may be broken off and even the die-plates may become damaged. Therefore, the strip length must be so determined that at the end there will be no partial holes cut.

There are warehouses that supply stock in sheared strips or unit pieces or round disks, etc. which may offer economic and manufacturing advantages with a better overall efficiency as a result.

Stock Inspection

New stock after reception must be checked first for correct dimensions, and afterward visually checked for possible defects.

In every case—whether the stock is in the form of strips, coils, full sheets, unit pieces, recuperated scrap, etc.—the material must be clean and free from oxidation, loose scale, corrosion, etc., otherwise the cutting edges of the dies will be worn too quickly. Consequently, the production performance of the tool becomes too low. If necessary, the stock must be submitted to some proper, preliminary cleaning process.

Other defects that must be detected before the stock is delivered for production are nonuniform quality, impurities, unsatisfactory surface finish, too great hardness (age hardening), waviness, bumps, slivers and laminations, pits and scratches, and bruises and dents. All these unsatisfactory features must be discovered before stamping, and questionable stock should be set aside, otherwise, too much labor, time, and money will be spent for subsequent corrective operations and for sorting out spoiled or unsatisfactory workpieces.

Finally, the stock must be straight and flat. Wrinkled and/or corrugated strips or coils often cause dimensional inaccuracies in the stampings, breakage of die members, and other types of problems.

Testing

If the material to be stamped does not comply with the requirements demanded by the stamping or drawing process, then there are problems, chiefly a very high percentage of rejects. Especially in the case of severely formed (drawn, embossed, coined, etc.) objects, the percentage of rejects depends on the characteristics of the stock, chiefly its ductility and formability, hardness, tensile and shear strength, surface finish, and springback index. Therefore, several testing methods have been devised to determine the properties of sheet metal. Every new batch of stock received from the supplier should be subjected to routine specimen checks before the whole batch is released for production. It is best if this inspection and related testing can be performed at the supplier's plant.

A Simple Ductility Testing Method for Sheet Metal

If you do not have some of the expensive devices for determining the ductility (ability to be formed easily) of metal sheets, you may test a specimen without any equipment, as has been proven repeatedly with success in actual practice.

Take a small (rectangular) piece of metal sheet and perform the following simple operations:

a. For light duty (simple bending, with comparatively large bending radius): simply fold the piece through 180° while interposing a piece of the same metal sheet (Fig. 5-1).

b. For medium duty (compound bending and/or small bending radius; low drawing) make the 180° fold without a spacer (Fig. 5-2).

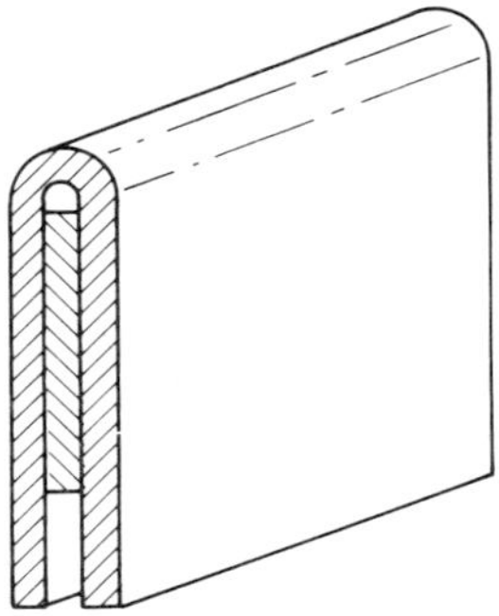

Fig. 5-1. 180° folding around a stock spacer made of the same material as the sheet.

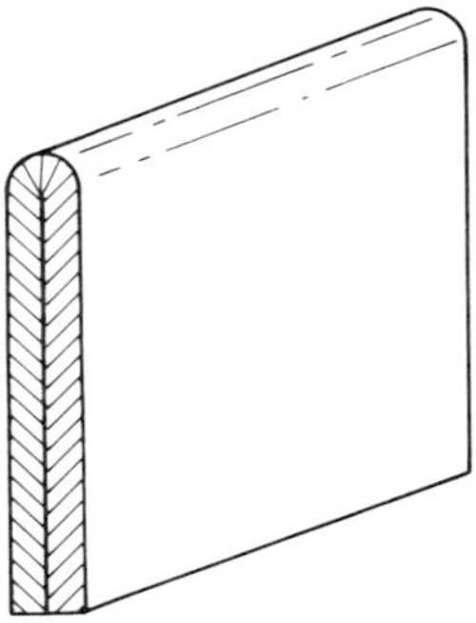

Fig. 5-2. 180° folding without spacer.

c. For heavy duty (deep drawing and/or profound deformations caused by deep embossing, etc.) use "handkerchief" folding. This consists (Fig. 5-3) of folding the sheet according to item (b) and then folding it again at 90° from the first bend.

In all these tests, the appearance of the metal sheet in the folding zone gives a good indication of its properties, i.e., its suitability for the forming processes to which it must be subjected. For greater accuracy, make the same test in order to find the presence or lack of cracks, tearings, orange-peel aspect, etc., simultaneously with other sheets of known quality, and compare the appearance of the respective folding zones.

Another analogous procedure of ductility testing consists of putting the specimen in a mechanic's vise and making a bend with a very small bending radius, then straightening and bending it again in the opposite direction (at 180° from the first bend), repeating this alternate bending until the sheet metal piece breaks. The number of bendings until rupture—by comparison with sheets of known quality—will give a good idea of the behavior of the metal being tested.

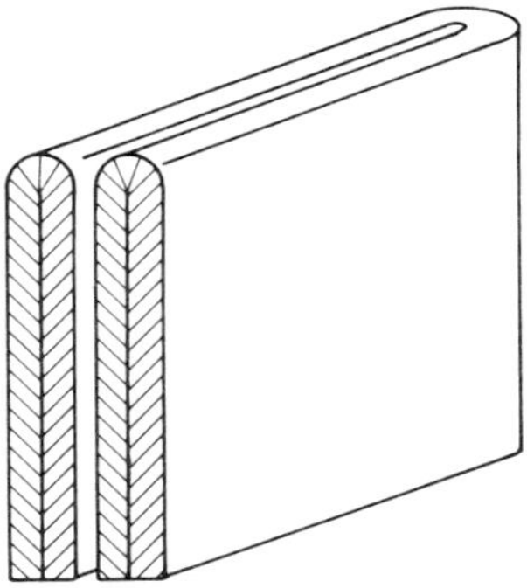

Fig. 5-3. Double 180° folding.

Materials Used as Stock for Stampings

Here follows a brief list of the usual and special materials which are employed as raw materials for stampings. It is not intended—nor is it within the scope of this book—to dwell too long on the subject, but only to give an idea about what kinds of material are used in pressworking. (As was previously mentioned, the selection of stock for a given stamping is a matter of component design. The metal stamper simply follows the designer's instructions and specifications.)

Essentially, every material that can be produced in the shape of sheet or strip can be worked with press tools, at least by the simpler operations. Excluded are only the brittle materials, such as cast iron, glass, and similar stock.

Commercially available materials for stampings may be divided into three groups: ferrous metals, nonferrous metals, and nonmetallic materials.

Ferrous Metals

As the name implies all alloys whose chief constituent element is iron belong to this group.

If no special requirements exist, the first choice for a stamping is low-carbon cold-rolled (or hot-rolled) steel (up to about 1/16 in., usually, for CRS; higher gages, usually, for HRS). There is no better or more economical buy for a stamping than this material. (It is estimated that about 50% of all stampings are made from SAE 1010 steel.)

To meet special specifications of strength and resistance to particularly adverse conditions (oxidation, corrosion, heat, cold), the steel must be a real *alloy* through the addition of certain elements. So, therefore, we may choose among high-carbon steel, alloy steels, silicon steels, stainless steels, heat-resisting steels, cold-resisting steels, etc.

Alloy steels may have any or a combination of the following additional elements (in addition to carbon): aluminum, boron, chromium, columbium, manganese, molybdenum, nickel, silicon, titanium, tungsten, and vanadium.

Nonferrous Metals

Obviously, to this group belong elements and alloys free of iron or where iron is not a predominant ingredient.

Aluminum is used in a wide range of alloys and tempers, and so are copper and its alloys—brass, beryllium copper, phosphor bronze, cupronickel, etc.—magnesium and its alloys; zinc and its alloys; titanium and its alloys; nickel and its alloys—nickel silver, monel metal, etc.—lead and its alloys (chiefly for impact extrusions); and tin. This group also includes the rare metals—zirconium, tantalum, columbium, tungsten, molybdenum, vanadium, cobalt, hafnium—and the precious metals—gold, silver, platinum, palladium, and their alloys.

Also, *prefinished metal sheets* should be mentioned. There are many possibilities for puchasing metal sheets (especially low-carbon steel and aluminum) with some protective or decorative coating. They are popular with many designers because of the advantages which they offer. Typical examples are galvanized steel (CRS zinc coated by hot dip process, sometimes tin coated, which is called tin-plate); galvannealed steel (galvanized sheets heat treated after coating thus producing a zinc–iron alloy); electrogalvanized steel (electrolytic zinc-coated CRS sheets); phosphatized galvanized sheets (chemically processed in order to receive paint or primer); and terne sheets (CRS coated with an alloy of 85% lead and 15% tin).

Aluminum-coated CRS sheets, coated by immersion in molten aluminum, are also available and are (1) heat resistant and (2) corrosion resistant.

Clad metal consists of a sheet metal base intimately covered by a layer of a different kind of metal or alloy. Usual combinations are steel plus stainless steel, nickel, nickel alloys, copper, or silver; copper plus stainless steel or platinum; brass plus gold, silver, platinum or palladium; aluminum plus aluminum alloys chemically different from the base metal or core.

Prepainted metal sheets are available in many colors; preplated by chromium, nickel, cadmium, etc.; lithographed; with plastic coatings on steel and aluminum; and with thin paper sheet for protection of highly finished, polished surfaces.

Finally, cold-rolled steel may be purchased with a lubricating coating applied at the mill during the manufacturing process.

Nonmetallic Materials

The possibilities and economic convenience of using press tools for the production of parts made from nonmetallic materials are too often overlooked.

Nonmetallic materials may be classified into two categories: natural and artificial. To the first belong: leather, hard rubber, soft rubber, mica, asbestos, felt, fabric, etc. To the second belong: plastics, celluloid, vulcanized sheet fiber, masonite, linoleum, paper, cardboard, etc.

According to the consistency and hardness or several kinds of these materials, dies of standard design are used; for softer stock, so-called dinking dies or hollow dies (similar to cookie cutters) are preferred.

Note: The majority of metals (at least those most widely employed for stampings) are commercialy available both as sheets and coils. The nonmetallic materials are usually sold as sheets only.

The thickness of metal stampings varies widely, from foils of 0.003 inch up to plate of 0.5 inch or even more. However, most stampings are made from medium thicknesses ranging between 0.031 and 0.125 inch. According to an arbitrary classification, it is usual to consider as light gage sheets those up to 0.031 inch, medium gage, from 0.031 and 0.109 inch, and heavy gage, over 0.109 inch (over 0.250 inch is called plate).

CHAPTER SIX

Stamping Techniques

Production Planning of Metal Stampings

The production of metal stampings—like the industrial production of any metallic or nonmetallic part—is based on a rational blending of five basic elements: *methods* (production organization or preparation or planning); *machines* (chiefly presses in case of stampings); *means* (press tools, press attachments, building, material handling equipment, etc.); *material* (stock); and, personnel (press operators, die setters, toolmakers, floormen, etc.).

To link and "tune" these factors correctly in such a way that, with the least input, the required quantity and quality of parts are produced in the scheduled time is the task for the people responsible for production.

One of the key persons in the production of metal stampings is the Production or Process Engineer. After receiving the order from management to start the production of a given product, the Production Engineer is responsible for the entire production operation. After a component has been designed, tested, and approved for production, the Production Engineer must proceed with the planning and organization of the production process. For such a task, a good slogan is "First plan production; then produce according to plan." Too often, people do not follow this advice but take shortcuts and the consequences may be quite serious.

The planning of production jobs, as specified in this chapter is, of course, different from part to part, from plant to plant, from company to company. However, there are certain principles that are universally accepted and followed.

The planning of an operation line-up consists of devising the simplest, most efficient, most economical, and fastest way to produce a workpiece, in all its details (every phase or step of production). In other words, the operation line-up comprises the whole series of technological steps needed to transform a piece of sheet metal into a stamping.

As mentioned above, there are no hard and fast rules for the elaboration of operation line-ups. The suggestions given in this chapter are only guidelines; they are of a general character and must be taken in consideration accordingly. At any rate, the best technology must be devised taking into account past similar experiences of the shop and the available production equipment.

Also, the production planning should not be exaggerated in every smallest detail. Some flexibility should be left to the experience of the shop people who are responsible for actually carrying out the production plan.

The first step consists of getting acquainted with the job. All the available engineering information (blueprints, samples, specifications about stock, dimensional tolerances, surface finish, etc.) is gathered and studied. Next, the proper operational cycle is established. (Usually this has already been done by the tool engineering people; the production engineer only has to check it and modify it, if and when necessary or convenient.) Afterward, the proper equipment is selected. Finally, the job information (called operational line-up, process sheet, operations sheet, production or work flow chart, etc.) is prepared. Be it a short run job or a mass-production (long-run) job, be it a small, medium, or large shop, in any case the pressroom people must be told in advance, accurately and in detail what to do, how to do it, and (possibly also) when to do it. Therefore, the operational line-up should contain all the pertinent information needed by the production people. The structure of process sheets varies, obviously, from shop to shop (and even from one job type to another). However, there are certain basic points which should always be contained in production flow charts. The chief basic items are specified and listed in Chart 6-1.

Chart 6-1: Operational Line-Up (Standard Process Sheet Chart)

1. Department name and number.
2. Individual number and total number of sheets (if more than one sheet).
3. Issue date; initial or name of process engineer.
4. Job identification data; number or symbol of the workpiece, its definition, and description (name); number of the corresponding blueprint, etc.
5. Stock: kind of metal; type (identification symbols, chemical analysis, tradenames); quality (commercial, drawing, deep drawing, fully aluminum killed, etc.); thickness, width, length, and other dimensions; hardness; surface finish conditions (if special), etc.
6. Shape and dimensions of the original material: whole sheet, sheared strip, coil, unit piece, blank, preprocessed stamping, etc.
7. Quantities of parts to be produced: total and partial (for each run or lot); production rate (quantities per time unit—hour, day, week.)
8. Time standards for die-setting, actual operation, handling, etc.; incentives.
9. Standard weight of raw material, e.g., pounds per 1000 units and total weight for the whole run or lot.
10. Identify number, name, and quantity of operations and their proper sequence, and not only press operations, but also any other manufacturing and/or finishing process. Specify in detail every pertinent characteristic and data of each operation: For example, Operator 3—PUNCH FOUR (4) MOUNTING HOLES 0.134 IN. DIAM., OR BEND AT 90° ANGLE. Provide, possibly, sketches of the part after each operation.
11. Tooling. Die(s) required, identification numbers, coding symbols, location in the tool storage room or tool crib, and any other vital, pertinent information.
12. Specify whether one or more operations are performed at any given press stroke (multiplex and/or progressive and/or compound dies).
13. Specify whether one or more workpieces are produced at any given press

Chart 6-1 (Continued)

stroke (if not specified, it is understood that only one stamping is produced at every press stroke).

14. If it applies, specify lamination grain direction.
15. If it applies, specify burr side.
16. Specify maximum admissible burr height, if it has some influence on stamping performance.
17. Number of press attendants needed. If not otherwise specified, it is understood that only one operator is needed.
18. Presses assiged for the operation: type, size, identification number, location, etc. Sometimes, also, an alternate choice is given, in case the originally specified press is tied up with some other job or is damaged or otherwise unavailable.
19. The auxiliary equipment and press attachments needed for the operations: bolster plates, air cushions, feeding and unloading devices, safety devices, material-handling equipment (conveyors, chutes, containers—specify quantities per container). Accessories: gloves, aprons, tongs, face shields, towels, etc.
20. Include inspection plan. In case of unusual or complicated stampings and/or in case of close tolerances make a detailed set of procedures for the essential dimensions and characteristics of the stampings to indicate the critical points liable to give trouble and which must be checked more carefully; also, how to inspect them correctly, wnen to do it, and by what means. (Specify inspecting gages and functional inspecting fixtures, if they apply.)
21. Stock lubrication. Specify type of lubricant and application method.
22. Specify handling instructions. Types and kind of containers. Whether or not they are lined with soft fabric, newspaper, etc. If special care is needed, say so. Exclusion of shovels, or other special provisions, must be stated. (For example, "do not fill containers to full capacity" or "do not throw stampings").
23. Packing instructions (if special).
24. Special instructions or notes.
25. Planned date for production to start.

Note: Operator has to register on the chart start and finish of production run together with any unscheduled interruptions.

Equipment Utilization

The very first condition of an efficient and productive shop consists of utilizing existing equipment to the maximum. In other words, every effort must be directed toward the goal of keeping as short as possible and avoiding completely in ideal cases machine downtime and the consequent inactivity of operators.

Idle equipment costs money, the same as fully utilized operating equipment. There are fixed expenses which are incurred, regardless of whether or not the machines are busy producing profits. Such expenses are interest on capital investment, rental of floor space, insurance, taxes, depreciation, overhead, and, in many cases, also salaries.

The various phases of the different operational cycles of stampings that are simultaneously produced in a press shop must be properly scheduled by means of a master scheduling chart in such way as to get the maximum equipment utilization. In other words, high press occupation factors are achieved by special attention to the preparation of work schedules and to strict production control. In other words, provide proper availability of stock (or semiprocessed workpieces for second operation jobs) and the necessary production equipment, in the right place, in the right quantity, and at the right time. By right time is meant somewhat before it is needed, but not too early. The latter is also an inconvenience: the place becomes cluttered and the waiting material may disturb the regular flow of work.

Industrial Engineering

The industrial engineer has a very rewarding position in a metal-stamping shop. The fundamentals of his profession, such as work measurement, job evaluation, method improvements, plant layout, and economic evaluation, are the basic ingredients of productivity improvements. His recommendations, suggestions, and proposals, if backed by top management, almost always mean improvements in production efficiency.

The shop manager, the superintendent, and even the formen, are concerned with producing the expected amount of products on the scheduled dates; they are occupied by routine problems and do not have sufficient time or peace of mind for studying individual tasks for improvements. It is necessary to depend on the industrial engineer for this. The industrial engineer's main functions are:

Establishing production standards and incentives

Work simplification

General troubleshooting

Standards

Production standards are established so that shop management has definite production requirements per employee; production rates are then a definite, measured quantity, and not a subject of argument.

Standards are established by time studies, which are used directly or by conversion into elemental standards. Standards include allowances, such as (average values) 5% for personal reasons, 2% for unavoidable delays, 5% for rest periods, and cleanup time, total 12%.

Wage Incentives

The pay rates established are those that an average, normal operator, skilled in the work and physically and mentally suited for it, can maintain steadily, day in and day out without injury to his health. The operator who has above-average skill or gives greater than the average effort should be rewarded on an incentive basis for his production.

Improving Stamping Productivity

The organization of the actual production offers excellent opportunities for production efficiency improvements without the need for large investments in new equipment. Usually, for a start, it is enough to make a rational rearrangement of the existing equipment, a stricter and more systematic scheduling of the production, and proper training of the workers. Essentially, one must try to shorten the total time needed for the production of a given workpiece.

The time needed to manufacture a certain product is the sum of three elements: (1) the time needed for the actual production process; (2) the time needed for handling and transporting the stock, the product, and the scrap; and (3) idle time.

Basically, production should be so arranged that the presses operate, if possible, all the time. (Do not forget that a press that is *running* is not always *operating*, i.e., producing.) This is achieved by having every auxiliary job reduced to a reasonable minimum and performed by auxiliary people, with the timing of all these operations being arranged in accordance with the press operator's job.

Also, the main time portion (i.e., actual press-operation time) may be reduced by means of optimum tooling, provided that there is economical justification for the corresponding investment. (This is usually the case where there is mass production of stampings.)

A few of these methods are listed below:

a. Progressive-type and transfer dies eliminate totally or, at least, substantially the need for costly and time-consuming secondary operations and interoperational handling.

b. Multiplex dies (producing more than one stamping at each press stroke) save considerable production time, and sometimes give spectacular results.

c. Automatic feeding, ejecting, and occasionally, stacking devices considerably shorten the duration of the operational cycle (in comparison with manual feeding and manual unloading) thus increasing the production rate or hourly output of a press.

Further productivity improvements may be achieved by work simplification and the application of some practical suggestions based on actual practical experience.

Work Simplification

Basic Concepts

Work simplification is derived from the time-honored "time-and-motion study." Work simplification means essentially employing techniques that enable a given manual job to be performed in the easiest and most economical way. In other words, it aims to eliminate or reduce fatigue and to lighten the manual worker's task.

In the majority of cases, work simplification is realized by comparatively simple means and little or no investment for new equipment.

The tangible results of work simplification in the improvement of production efficience are twofold: direct and indirect. There are direct results because the actual working time is shorter, i.e., during a given time period more workpieces are produced (there is a production rate increase, i.e., more operations are performed by the same labor force and the same equipment). The indirect results manifest themselves through the reduction of worker fatigue.

Note. Obviously, in case of fully automatic production (and in many cases of semiautomatic feeding), work simplification has little or no place, because every bit of time saving has already been done during the study of the tooling and equipment. However, in the case of manual feeding of presses, work simplification is extremely attractive and useful, because of its influence on production efficiency.

The goals of work simplification are not reached suddenly by bright inspiration or by chance; they are the products of thorough scientific studies and investigations of every job and of every phase of every job, in order to see whether or not some motion may be simplified or reduced or even eliminated, i.e., to determine where human effort and time are wasted unnecessarily.

Of course, every shop, every job, and—up to a certain extent—even every individual worker is a different proposition. However, there are several basic concepts and practical data which have universal application, so that everyone may take advantage of them. This section will deal simply with these universal features.

Application of Work Simplification

Since every job may be performed with high, medium, or low efficiency, it is always worthwhile to analyze it in order to see whether or not it may be improved. Work simplification may be applied to any job or complex of jobs, or to any phase of a given job, independently of the other phases. In any case, the best approach to job improvement consists of following the stages of the procedure outlined in Chart 6-2.

Chart 6-2: Work Simplification Procedure

1. Select a job to be improved and define it.
2. Analze the job, breaking it down into its most-detailed elements.
3. Obtain and register every relevant fact of the existing method, referring to each detail of item 2.
4. Study systematically—in correct sequence—and critically every detail of item 3, so that every imperfection, even the smallest one, is clearly defined.
5. After the analysis comes the synthesis, i.e., plan and develop an improved (easier, more practical, more efficient, more economical) method.
6. If necessary, properly train the operators in the new method.
7. Test the new method in actual practice or, at least, in conditions resembling actual production and if everything is all right, install it as a standard procedure.
8. Maintain the new method through continuous, periodic, routine observation, inspection, and checking.
9. Never consider a new, improved method as inflexibly definitive. Recheck it and its usefulness at reasonable intervals (3–6 months or so) in order to see whether or not some new process or means could in turn improve or replace or eliminate it.

Means and Methods

For both the analysis and the synthesis (items 3–5 of Chart 6-2) it is necessary to proceed methodically, trying not to overlook any detail. Here follows a good pattern, which is of considerable help in these studies.

I. Challenge the need for the operation. Often it may be eliminated or combined with some other operation. In some cases an inspection may be combined with an operation. Sometimes the part produced by the operation may be purchased at lower costs.

II. Equipment, tooling, feeding, and ejecting must conform with the best practice suitable for the job to be performed, and every unit of production equipment must be kept in the best operating condition.

Special attention must be given to proper press speeds and feeds. Always use the maximum speeds compatible with the job and the operator's fatigue limits. Consider the advantages of variable speed drives.

III. Working methods.

a. Consider whether or not the production method can be changed, for example, substituting drilling by punching, reaming by shaving, etc.

b. Provide easy and comfortable handling of presses, dies, stock, parts, scrap, etc. Everything that is necessary to perform the job must be kept in an orderly, prepositioned fashion (good housekeeping!) within easy reach of the operator and correctly placed with respect to body position. In this way the operator's energy is not wasted with unnecessary movements. Scrap and product should be disposed of, possibly automatically. If it is not possible to do so automatically, then by proper drop chutes (gravity is the easiest method for this purpose).

c. Consider the possibility that the operator should do some useful work during the running time of the machine, instead of having to wait until the press is fully stopped. For example, during the ram up-stroke the operator may prepare the next blank to be fed to the press or he may stack the parts produced by his press, so that the workpieces will be presented in a preestablished and correct position for the next operation (see below).

d. The operator should not idle his machine or stop his regular routine in order to get fresh supplies of stock or to deliver the produced parts.

IV. Motion economy. This detail is one of the most interesting and valuable means for improving production efficiency. It is divided into parts each with several factors that may be applied either separately or concurrently.

a. Reduce the number of motions. Every movement should perform something useful. Habitually futile movements must be eliminated. As many motions as possible should be productive (not used simply to transport materials, i.e., microhandling). Remove obstacles from the movement path. In order to avoid waste motions, keep parts clean and use improved containers that facilitate grasping.

b. Keep movement extent limited to a minimum. The distances to be covered by any part of the operator's body should be kept as short as possible. This means (1) everything within easy reach; (2) proper work height—no need for bending or stretching; (3) disposal and fetching should be in the same general direction; (4) momentum should be utilized as a worker aid; and (5) equipment should require minimum change of body position—simply and easily reachable mechanical control devices should be provided.

c. Make movements simple. Carry out the job with the lowest possible class of motion, compatible with satisfactory working conditions. This means that fingers should be preferred to hands; hands to arms; arms to legs; legs to trunk. Obviously straight movements should be employed to reach from one point to the other (simplest and shortest movement). However, if this is not possible, then it is preferred to effect continuous smooth curved motions instead of straight line motions involving sharp corner directional changes (zig-zag motion). Curvilinear movements are swifter, easier, and more accurate; they should be preferred to controlled or restricted motions.

d. Two-hand motions. Both hands should be utilized simultaneously for doing some useful work. They should move concurrently, i.e., start and perform, possibly, the same work and finish their movement at the same time. Hand movement must also be symmetrical about an immaginary line through the center of the operator's body. By means of such balanced motion the operator may easily maintain a correct posture and avoid fatigue. Two hands should not be idle at the same time, except during rest periods. Whenever possible, hands should be relieved of work that does not need the dexterity of hands, i.e., work that can be done satisfactorily by the feet (pedals).

e. Movements should be natural. It is essential to plan for the easiest natural movements of the body, with a minimum of tension in the body members. Loads should be distributed according to the capacity of the corresponding body members. For best results, in order that one or a series of movements become automatic, habitual, proper, easy, and natural, rhythmical cycles should be established so that the parts of the body can move smoothly and freely.

f. Holding devices. Hands and fingers should not be used in place of work-holding jigs or fixtures (except in very special cases); hands should be reserved for more important productive functions. Workholding should be effected by means of the correct vises, fixtures, jigs, etc.

Operational Cycle of Stamping Processes

In applying work simplification specifically to stamping techniques, one must examine the individual phases of a complete operational cycle.

The operational sequence of a typical single-stroke second operation cycle with fully manual feeding and unloading and with hand controls is composed of a number of phases as listed in Chart 6-3.

This operational sequence list is the most complete and complicated one. In other cases the sequence is simpler. For example, in the case of single-stroke first operation (with hand feeding) the press is actuated usually by foot pedal treadle and the operator's hands do not leave the strip (or coil) between operations. Consequently, phases (1)–(8) are converted into one movement: "advance" or "feeding." In addition, phases (12)–(15) become inherently automatic, the blank being evacuated by gravity. Consequently, the number of steps to be performed by the operator is reduced practically to three, viz., feeding, tripping the press, and the ram stroke. With fully automatic feeding (whether first operation or second operation) practically all the steps are eliminated.

Let us consider now the chief phases in order to find out how they can be shortened to obtain higher production rates.

1. For easy reaching of the blanks (or strips) they must always be located in a suitable position, at the proper height and correct distance from the operator in such

Chart 6-3: Complete Sequence of a Press Operation Cycle

1. Move hand to part stack
2. Grasp part
3. Lift part
4. Orient part
5. Carry part to tool
6. Introduce part into tool
7. Release part in tool
8. Position part in tool
9. Lubricate part periodically
10. Move hand to controls (if not foot controls)
11. Trip press
12. Wait for full complete press-ram stroke (if necessary, holding the part during the ram stroke)
13. Move hand to part
14. Grasp part
15. Remove finished part
16. Inspect part periodically
17. Carry part to tote box
18. Put part in tote box (stack, if necessary)
19. Move hand to scrap
20. Grasp scrap
21. Remove scrap
22. Carry scrap to bin
23. Put scrap in bin and release

a way that no excessive stretching or bodily movement is required (which would cause premature fatigue).

2–8. The introduction of the blanks into the dies should be facilitated by means of proper chutes or other means.

10. Hand travel. In single-stroke technique generally too much time is lost, usually by long travel of the hands to and from the press controls. (These should be at an easy and comfortable reach.) Sometimes this can be remedied by making the actuating buttons movable (located in panels which are adjustable both vertically and laterally) in order to locate them so that there will always be the shortest hand travel for the actuation.

In other instances, a valuable aid is provided by electronic safety devices by means of which hand travel to and from the actuating (control) elements may sometimes be eliminated completely. In such cases the press is automatically tripped as soon as the operator's hands are removed from the light curtain. Of course, there must be assurance (by means of mechanical, electrical, or other interlocking devices) that the workpiece to be processed is in its correct position. An analogous result can be obtained with slides or other feeding devices where the interlocking mechanism, which is provided to prevent the descent of the ram if the workpiece to be stamped is not in its proper place, automatically actuates the tripping of the press. Also, the movable safety devices may be combined with automatic tripping of the press. In all these cases there is a considerable production rate increase because in the meantime the next workpiece may be prepared (taken in hand to be fed for the next operation cycle).

11. Tripping the press. With single-stroke technique, the starting phase must be taken into account; it is a mistake to neglect it even though it is brief. By actual tests it has been found that it takes about 0.1–0.3 sec, depending on the design and conditions of the press and of the clutch. This phase, i.e., the time period which elapses between the pressing of the actuating elements and the actual start of the ram, is due to the inertia of the press, which is unavoidable.

It is not correct to consider the "speed" (i.e., spm) of a press only in terms of continuous running. If it is to operate in single-stroke stamping, then the starting phase and other secondary factors must also be taken into account.

The production rate of presses with short strokes (for cutting operations with the single-stroke technique) may be considerably increased by changing the clutch from the pin-type or key-type positive clutch to the disc-type air friction clutch.

Occasionally the actuation is performed with pedals. This may be done only if there is no hazard for the hands, i.e., the tools are perfectly "safe" or "accident-proof." (See Chapter 13.) With foot actuation, a notable production increase is possible because the hands may be usefully occupied otherwise (preparing the job for the next cycle, etc.)

At any rate, it is indispensable that the actuation of the press should be easy and effortless, in order not to fatigue the operator unnecessarily (for example; electrical microswitches which actuate the tripping control are excellent for this purpose). Light actuation also contributes directly to an increase in production efficiency.

It is important to instruct the operator with respect to correct clutch actuation: it must be done quickly and positively as with automobiles having standard transmissions, either "in" or "out," never in an intermediate, indeterminate position, because such handling causes premature wear of the clutch mechanism parts. (This observation refers to mechancial, dog-type clutches.)

13–23. Removal of workpieces and of scrap should, if possible, always be made automatically. Thus, these steps are eliminated. If part and scrap removal are performed manually, provide proper chutes or place the containers within easy reach of the operator, so that this activity can be performed with minimum effort and quite rapidly.

Remarks

Work simplification, luckily, does not require complicated, sophisticated equipment and the institution of large programs. It is quite possible—and even recommended—to start with a very modest program involving one or two tasks, enlarging it as results may indicate.

Another fact also must be taken into consideration, in this respect. Work simplification is essentially a productivity-conscious attitude toward work and requires a proper approach to the problem. It may be achieved only by harmonious relations between management and employees. Usually working people offer more or less open resistance to such improvements (either through fear of or reluctance to changes). So work simplification must be applied with diplomacy and patience, in order to get the full cooperation and collaboration of the respective workers, without which there is no possibility of achieving positive results.

Practical Hints for the Production of Metal Stampings

This section is a collection of reports of case histories and some useful practical information, taken from the author's actual experience. These are typical cases of stamping production techniques where the main goal is not only the solution of specific technical and other problems, but also the improvement of the production efficiency.

In the case of multistation progressive-type dies, especially when working with light stock, if the stock presented in the shape of coils is not maintained flat all the time, there is danger of corrugations, bumps, or kinks, with consequent inaccuracies in the workpieces (changed center distances result in shifted positions of the holes in the preliminary stations, and other problems). In order to prevent this, the strip skeleton should be re-reeled, thus maintaining tension on the stock and thus keeping it flat (Fig. 6-1).

Feeding of long and/or heavy strips (straight lengths) is troublesome, because of their overhanging weight, which makes their handling difficult. The strips are liable to bend downward before they enter the die and, if the strip skeleton is too large, also after they leave the tool. This may cause a problem in multistation progressive-type dies, where the center distances may be changed if the stock is not held perfectly straight and horizontal. In order to prevent this, the strip should be supported at the entry side of the die and the strip skeleton should be supported at the exit. For this purpose anything, within reason, will serve, even simple wood blocks (Fig. 6-2).

If the stock is sheared in strips in a squaring shear, the strip width is usually not very accurate. There is almost always some more or less appreciable variations in width. In the case of progressive-type dies, the width may be important, therefore, it must be checked before the press operation because there may be trouble if it is too narrow or too wide. Attach to the die shoe a hardened snap gage of the "go–no go" type and instruct the press operator to use only those strips which are of the correct width at *both* ends (Fig. 6-3).

In Fig. 6-4 is shown a die set with diagonally located leader pins. This die set was changed to another one having axially disposed guide posts. This change permitted the working process to be modified, i.e., the working positions were shifted from opposite to angular (Fig. 6-5) resulting in a much simpler and quicker operation.

In another case the press was equipped for a second operation punching (of a drawn shell), with two-hand controls, for safety. In the original design (Fig. 6-6) there was a single tool; the workpiece to be punched was put in place with the right hand and the finished product was removed from the die with the left hand. After modification (Fig. 6-7) two identical tools were provided so that both the right and the left hands could each put a blank simultaneously in place and then, also remove the finished parts at the same time. Thus, symmetrical movements have been made possible. Additional advantages are easier working, less fatigue, and higher productivity.

Positioning for forming (bending) is best made with internal gaging where pilot pins located in the pad or in the forming punch or in the blankholder engage corresponding holes in the blanks. However, this method requires close attention from the operator, some fumbling always results, and, thus, the operation becomes slower (Fig. 6-8). It is much easier and faster to introduce flat blanks into "nests," which are external locations near the outer contour of the blanks (Fig. 6-9). However, this method is not always accurate enough. The best solution is to put an open "fence"

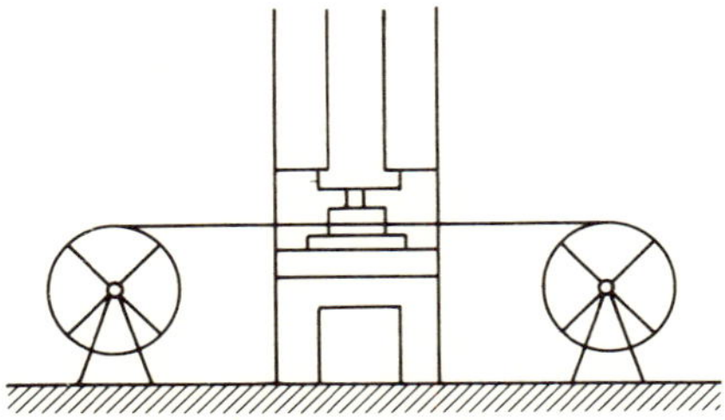

Fig. 6-1. Rewinding strip skeleton.

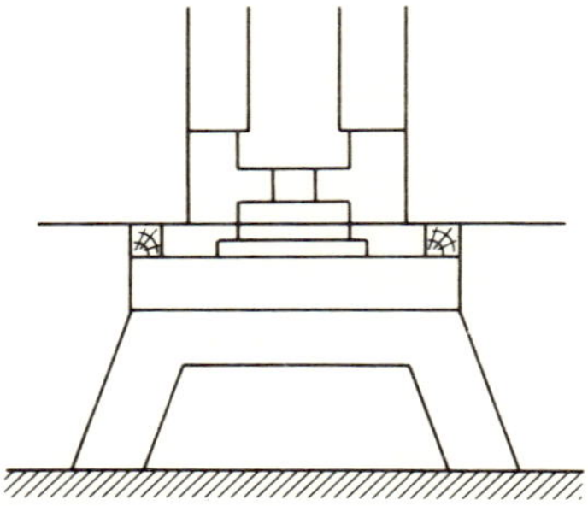

Fig. 6-2. Providing support for long strips.

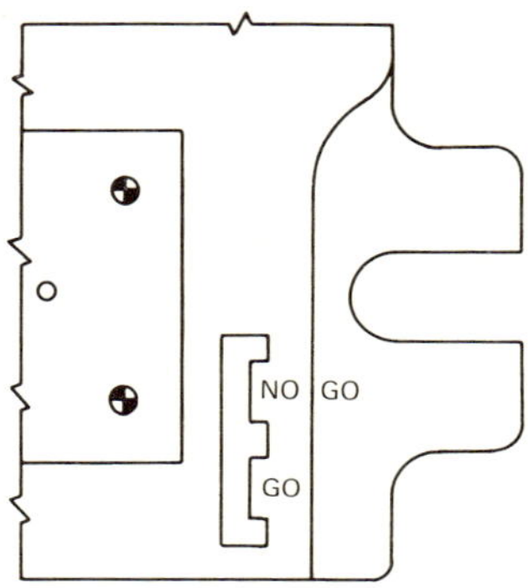

Fig. 6-3. Checking sheared strip width with "go–no go" gage.

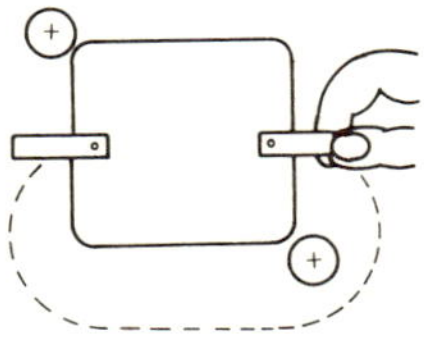

Fig. 6-4. Diagonal leader pins slow down second operations.

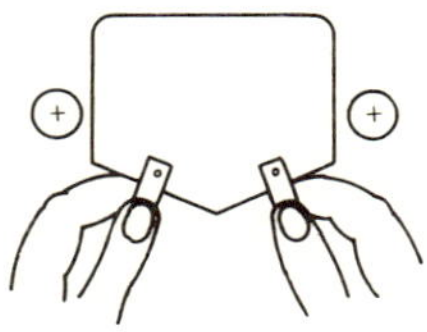

Fig. 6-5. Axial guide posts improve productivity.

Fig. 6-6. Single tooling gives low productivity.

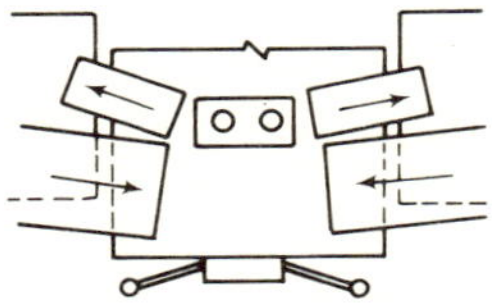

Fig. 6-7. Twin tooling increases product output.

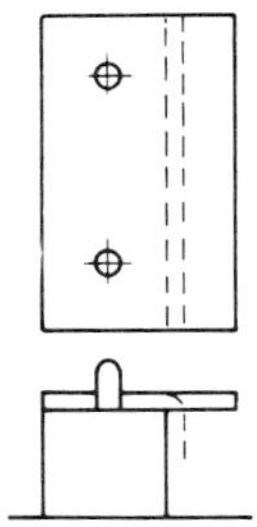

Fig. 6-8. Pilot-pin location is accurate, but the operation is slow.

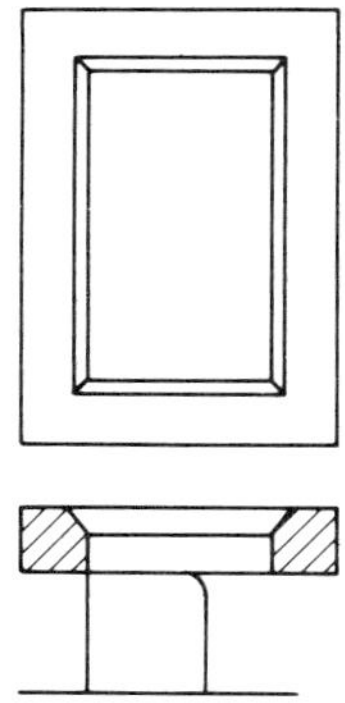

Fig. 6-9. Open nest gives higher production, but lower accuracy.

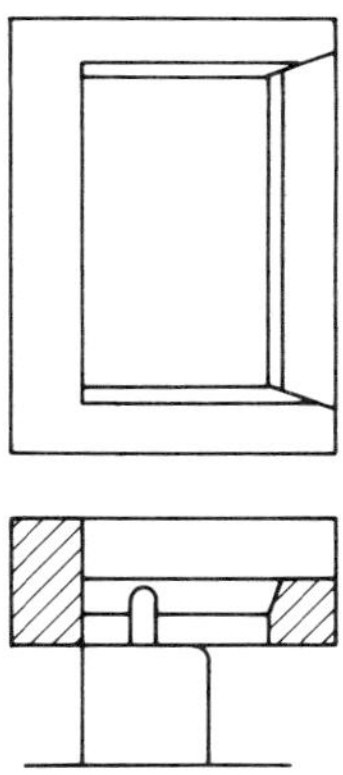

Fig. 6-10. Best solution is open nest combined with pilot pins.

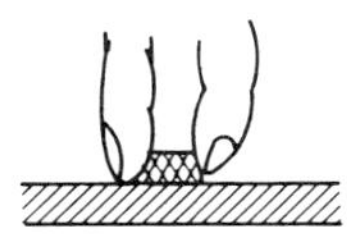

Fig. 6-11. Do not use a hard surface for picking up small parts.

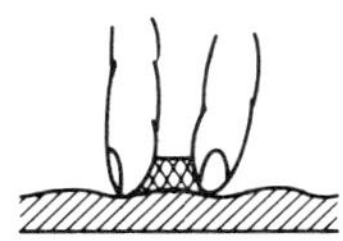

Fig. 6-12. Resilient pads make picking up easy.

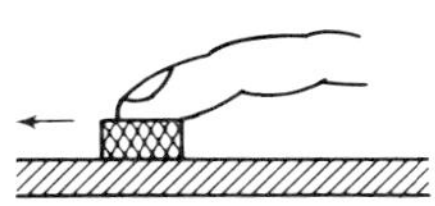

Fig. 6-13. Sliding or pushing is better than picking up small parts.

around the nest and also to use gaging pins (Fig. 6-10). In this way, the positioning of the blanks becomes extremely easy, quick, and accurate.

If small parts must be picked up by the operator, then do *not* use a hard table (Fig. 6-11), use resilient pads (Fig. 6-12). Speed increases of 100% and more have been achieved. An even better solution consists of sliding or pushing small parts rather than picking them up (Fig. 6-13).

A "table" or "tray" made from heavy gage sheet metal and mounted around a die so that the tray top surface is flush with the die plate top surface allows the press operator to put a handful of small workpieces to be processed (in a second operation die) on the table and to push them simply one by one into the die (Fig. 6-14).

When blanking brightly finished stock (aluminum, stainless steel, etc.) where the

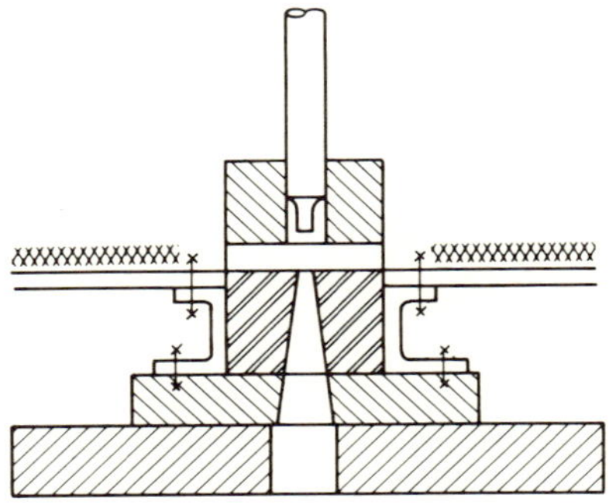

Fig. 6-14. Sheet metal tray around die helps second operation feeding.

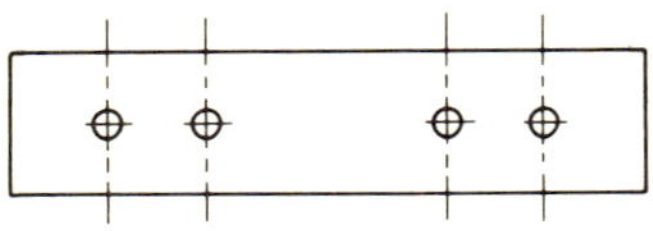

Fig. 6-15. Prepunched blank for easy forming.

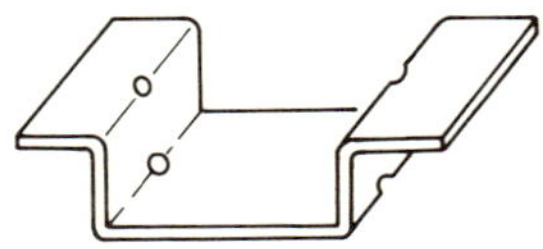

Fig. 6-16. Hot-formed stamping with prepunched holes.

high finish requires protection, cover those surfaces of the tool that come into contact with the stock (die plate, punch, stripper, pads, etc.) with cloth, excepting, of course, the actual cutting or forming edges. The cloth is best attached to the metallic surfaces by using shellac as an adhesive. Another similar suggestion for the same purpose consists of covering the material surface with oiled paper, pressure-sensitive paper, or some pealable plastic coating, in order to prevent any possible damage to the finished surfaces. The coating is left on the stock until the entire forming operation is finished, and, then, it is removed.

Difficult bending operations caused by too hard stock and/or too small bending radiuses and/or wrong burr side, etc., may be made easier by resorting to one of the following remedies:

Deburring. Eliminating the blanking burrs, before the bending operation, makes this operation much easier because the stiffness of the blank due to the burr is decreased. In short run jobs, the elimination of burrs and sharp edges can be done by filing, grinding, milling, etc., while in the case of regular mass production of stampings, it is much more economical to do it by tumbling.

Annealing. By annealing the blanks their ductility is greatly increased. However, annealing is a supplementary and somewhat expensive operation. (Often it requires another additional cleaning operation.) However, in extreme cases it represents the only solution for severe conditions.

Warm bending. Bending operations are usually performed at room temperature. They are "cold-forming" operations. However, in certain cases of extreme severity the blanks and even the tools must be heated, and the stampings are bent warm. Since an increase of the temperature increases the ductility of metals, and thus the bending action becomes much easier.

Hole punching. Quite often, small holes punched in the bending zone of flat blanks (Figs. 6-15 and 6-16) do not affect the function nor the strength of the parts.

On the other hand, such holes greatly facilitate the bending operation by requiring reduced pressures to obtain quite sharp bends.

Difficult drawing operations may be made easier in two ways. First, take into consideration the time factor. By reducing the ram speed (linear speed), there is more time for the plastic flow of metal and the rearrangement of the molecules is made more smoothly; consequently, there is less danger for the formation of wrinkles, tearing of shells, etc. In the case of very quick deformations, the metal has the tendency to crystallize and thus its ductility and even its strength is decreased.

Secondly, multidrawing can be done without annealing, or at least with much less annealing than is commonly needed, if the subsequent second operation redrawings are performed in quick succession. Thus, the metal has no time to cool; therefore, there is less, or no, work-hardening, i.e., formation of crystals, and, consequently, the metal retains its ductility much longer. About work-hardening during press operations there is little practical information available. However, it is known to take place after an hour or so. In actual practice, low-carbon steel has been drawn with the quick succession method as many as seven times without annealing between draws. If it is not possible to make the redrawings of shells in quick succession, then hold them in hot water [about 200°F (93°C)], between operations. In this way the metal will retain its ductility much longer. The quick succession method may be used either with multioperational dies or by providing a battery of single-operation presses, one for each redraw.

Sometimes roller leveling may solve difficult drawing or forming operation problems. By passing the cold-rolled stock through a proper roller–leveler, prior to entering the die, the stock is stretched, and, thus, the internal stresses are removed and, consequently, its ductility is increased considerably.

Finally, remember that the reverse drawing method allows easier operations than conventional drawing methods.

If lubricated blanks stick together, accidents, damage to tools, machines, wasted parts, etc., may occur, because if such blanks are used in a subsequent drawing or forming operation, it is often unavoidable that two blanks be fed into the tool. In order to prevent this, whenever the function of the component allows it, stamp in the blanks a small projection. This may be made with the blanking punch, itself. It is suggested that the embossing be located in an off-center position (Figs. 6-17 and 6-18).

It has been found empirically that increased cutting speeds (in the region of 1000 strokes per minute) give several advantages. First of all, of course, there is higher hourly output and consequently, better equipment utilization. In addition, there is improved quality; i.e., there is less tendency to burr formation.

When small metal or plastic workpieces are produced automatically in mass production by punch presses, it is customary that one operator attend to a battery of

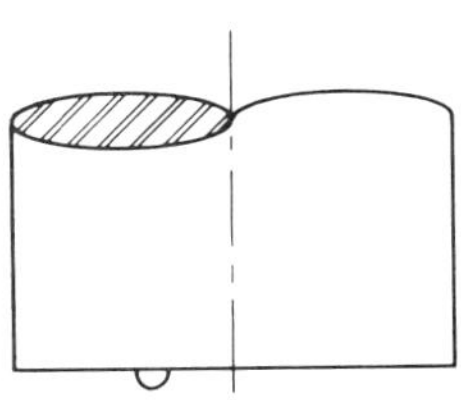

Fig. 6-17. Punch with pin for shell blanks.

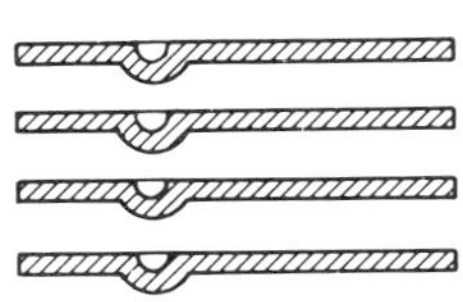

Fig. 6-18. Shell blanks with small protuberance for easy separation.

machines. He sets them up, starts them, feeds them, and checks their production periodically. This is satisfactory as long as there are no problems. However, when some problem does arise, the machine in question starts producing rejects. These parts get mixed up with the good ones, and it is necessary then to separate them by tedious, costly inspection.

There is an easy, inexpensive solution to this problem. Use the following method, which has given very satisfactory results in actual practice.

Every machine is equipped with a small tray which is made from perforated sheet metal. Thus, it is light and easy to handle and any lubricant may drain easily through the holes. This tray is put at the exit chute of each punch press. It catches the stampings as they are ejected from the machine. The operator checks his machines periodically at roughly constant intervals, inspecting the last piece that comes out of the press. If it is satisfactory, then the content of the tray is poured into the drum or container that stands by the press. If the last stamping is unsatisfactory, the operator stops the press, and only the small quantity of stampings that are in the tray must be inspected to separate the good pieces from the defective ones.

CHAPTER SEVEN

Working Conditions

As was mentioned in Chapter 6, the production of metal stampings consists of five basic factors: the method (organization of work), the element to be transformed (stock, raw material), the chief transforming element (machine, press), the auxiliary transforming elements (tools, devices), and the transformer element (worker, operator). Each element contributes to the success or failure of the production and has its own particular importance; only through complete and unhindered utilization of all these basic factors can the desired production efficiency be attained.

For press working, usually unskilled labor is employed. This is a drawback. Ideally, the operator of a press should be a comparatively well-educated person, with a full sense of responsibility and with a certain rudimentary technical knowledge. The operator should be generally familiar with the working principles of presses, dies and accessories, processes, the characteristics of stock, etc. In addition, the operator must be indoctrinated and trained for good housekeeping and safety (see Chapter 13).

The operator may contribute a great deal, directly and indirectly, to the improvement of production efficiency. The tool, press, feeding and unloading devices, accessories, stock, tool- and work-handling means, in short, the whole shop organization, create only a certain *possibility* for a given maximum production efficiency. It is the operator's task and duty to take advantage of this possibility and convert it into a *reality*. In other words, in the case of manual and semiautomatic feeding and ejection, it is the operator who actually governs the production rate, according to the speed with which he or she works. This is the direct aspect of his or her cooperation.

The indirect aspect is perhaps even more important since it contributes, paradoxically, more to the results. It consists of reducing work spoilage, rejects, and problems at large.

Physical Working Conditions

Adequate and proper working conditions and congenial surroundings contribute notably to employee morale and to production efficiency improvement. It should be kept in mind that the worker is first of all a human being, and only secondarily a producer, so his production efficiency depends in great measure on the environment where he works. Basically, people are happier and work better if surrounded by an attractive, clean, orderly environment. These conditions must be provided primarily by management; however, operators must also cooperate considerably in order to achieve the best results.

With the exception of a comparatively few, very modern companies, all existing industrial plants could do with some more or less substantial improvement in working conditions. The results would often be immediate, directly in increased productivity and indirectly in improvement of production efficiency. Even a minimum improvement of morale or of the psychological conditions of personnel often results in some productivity increase.

It is surprising how many details must be considered with respect to satisfactory physical working conditions. This chapter deals with them only briefly. The reader who is interested in the subject can readily get more information on the pertinent details and adapt them to the individual characteristics of his own plant.

Good Housekeeping

Everybody knows (or at least, is supposed to know) what "housekeeping" or "good housekeeping" means: cleanliness and orderliness. Only in a comparatively clean and uncluttered plant can satisfactory, high-quality work be performed safely and quickly.

Cleanliness

Cleanliness is the first essential condition for workers' health and for high-quality product manufacture. Cleanliness may be a matter of hard work, but, in general, it is not too dificult to achieve and usually is not expensive.

Both management and employees must do their share in order to achieve and keep a clean shop. Although the responsibility for proper plant cleanliness rests chiefly with management, each press operator must be made directly responsible for his own press, accessories, and working area.

Advantages gained by cleanliness and tidiness may be summed up as follows:

Clean, well-cared-for machinery is treated with greater respect by its operators.

First-class quality products are only possible in a clean plant.

Clean equipment is easier and less expensive to maintain.

Of course, the cleaning schedule varies substantially from shop to shop, but the basic questions remain: What to clean, when, how (with what), and who does it?

In order to draw up a schedule, a complete list must be made of all the items requiring cleaning—floors, walls, windows, skylights, ceilings, machines, furniture, lighting fixtures, blinds, piping, etc.—and time periods allotted for each, with preference given to those periods when equipment is not operating.

If the size and type of shop allows it, portable, power-driven, labor-saving cleaning devices, such as power sweepers, floor washers, polishers, and industrial vacuum cleaners, are a good investment.

In every plant a central cleaning department must be made responsible for carrying out the cleaning program. Once a plant is brought up to a certain level of cleanliness, the day-by-day effort required to maintain an orderly, clean workplace is negligible in comparison to the time and expense that would be required for massive cleanups at long and irregular intervals.

The consequences of neglect ramify as in the case of tote boxes. Dirty or oily transportation boxes soil the workpieces which they carry. When these unclean blanks are put into second-operation dies, the dies become dirty, easily scored, and worn leading to further spoiled work until the defect is discovered and the tool has to be reconditioned. All this means a reduction in production efficiency.

On the other hand, even if the soiled condition of the blanks *is* noticed by the operator, causing him to stop and clean every blank before introducing it into the next die, there is still an unfortunate production rate decrease. (A good way to avoid dirty, oily tote boxes consists of lining them with cardboard, heavy paper, or old newspapers, which may then be changed, as necessary.)

In first-operation cutting dies, the cleanliness of the strips (coils) is just as important as in secondary operations. Dirt (e.g., deposited cement floor dust) always acts as an abrasive, leading to premature wear of the cutting edges of die plate and punches.

Orderliness

The concept of orderliness may be broadly expressed as: "A place for everything and everything in its place." In addition, orderliness involves fundamental principles of efficiency, i.e., avoiding unnecessary work or spoilage by the fullest utilization of time, labor, means, space, and equipment.

Good housekeeping in the press shop requires planning or foresight. In every press shop with a proper plant layout the storage areas (for raw materials, parts-in-process, finished parts, scrap, supplies, etc.) are separated from the working areas and traffic lanes (aisles). These traffic lanes (also doorways, stairways, fire escapes) must always be kept free of obstructions to allow passage. Such aisles and traffic lanes should be clearly defined on the floor, with yellow paint or tape.

Keeping the working areas in order means that nothing should be left on the floor, under machines, under pieces of equipment, or on window sills. Cleaning materials should be accessible. It is good practice to provide boxes for cleaning rags, fastened to the press frame, in easy reach of the operator.

The diesetter is one of the key men for achieving good housekeeping. With his daily example of orderliness and cleanliness (by keeping the presses and surroundings well ordered) and by pertinent instructions given to the press operators, he contributes considerably to the success of good housekeeping programs.

It is the press operator's duty to straighten up his working place at the end of each run and/or each shift. He must retrieve any good parts which may have fallen on the floor, clean them, and put them into the appropriate bins or drums. Any loose scrap must be collected and put into the scrap bin. The floor around the press must at all times be kept free of oil, grease, water, chemicals, and litter.

Tools and accessories not in use must be returned to racks or tool cribs. A central area should be designated for company-owned tools and supplies, not immediately required.

At the end of each shift or run, all containers with completed parts, located on skids or pallets, must be properly tagged and moved to shipping or storage areas. Here they must be kept off the floor to facilitate sweeping. Such tote boxes or drums or other containers should never protrude into the aisles.

Proper scrap handling demands that at the completion of each shift, all scrap be removed from around a press and all scrap containers be deposited at the scrap collection area. It is even better to remove scrap several times per shift.

Rejects also must be collected and stored in a designated area for earliest return to the supplier in the case of unsatisfactory raw material.

All refuse should be placed in convenient, strategically located, and properly identified containers (possibly of self-closing design) separating waste, rags, paper, and scrap; in addition, sand-filled containers should be provided as ashtrays, and proper containers should be provided for empty soft drink bottles. All of these must be emptied on a regular schedule.

Good housekeeping is a matter of mutual cooperation between management and employees. The operators must be made aware of the personal advantages to be gained through good housekeeping, which include better working conditions and a safer shop.

Lighting

Adequate illumination of the working areas contributes significantly to the well being of the workers and improves their productivity.

Additional benefits are:

Improved morale among employees at large (by helping to create agreeable working conditions), especially among older, experienced employees (whose efficiency is improved); consequently there is less labor turnover

Since poor illumination is a decidedly active contributing factor to accidents, good lighting means better safety conditions

Increased worker welfare in general; improved housekeeping, cleaner, and neater appearance of the shop

Satisfactory lighting conserves workers' good health by helping to prevent eye problems.

Each shop and factory presents different lighting problems; it is impossible to set hard and fast rules. However, there are several basic principles of good illumination which must be followed in every case.

Lighting must be considered and examined from two different, yet interrelated standpoints: day (natural) and night (artificial) illumination. In the great majority of cases there is only one shift, which may therefore not require artificial light (or possibly require it for a maximum of only a couple of hours a day in wintertime). Consequently, natural light has at least as much importance as artificial illumination. In fact, some experts consider that nothing can substitute efficiently for sunlight, not only because of its intensity, but chiefly because of its compatibility with human vision.

It has been found that in windowless factories (with 24-hour artificial illumination) the employees often work with lower efficiency. This is probably partly due to psychological causes: some people do not like to feel shut in, while others like to look out a window, once in a while, for a change of scene.

At any rate, every machine or working place should be placed so that the light coming either from windows or from skylights or from artificial sources should be abundant without being excessive and without creating glare, unpleasant reflections, or disturbing shadows.

The basic requirements for good illumination (chiefly with regard to artificial light) may be summarized as follows:

First of all, remember the basic rule: "Light the work and not the space"

Adequate light intensity should be based on the type of job to be performed, in order to reduce to a minimum—or if possible, entirely prevent—eye strain.

Take into account, if possible, the age of the workers involved. Older people need more intense light than younger people.

Light intensity should be as uniform as possible (maximum ratio 1 to 3), otherwise there is fatigue and eye strain for the worker. Also, too high a light intensity (over 1000 lux) often creates more harm than benefit as it increases fatigue.

The light should be properly diffused. Consequently, harsh shadows or semiobscure zones are minimized or totally eliminated, especially in the area of operation.

Prevent, absolutely, direct glare (positively no unreflected light in the direct field of vision of the worker) and indirect (reflected) glare. In fact, glare can cause eye strain, and it is responsible for many accidents.

The light must be steady and free from noticeable oscillations.

Correct lighting of press shops consists of a good general illumination combined with supplementary localized lighting of the workplace, especially where needed for safety reasons or better inspection.

It is good practice to provide large presses with adjustable or fixed lamps located on uprights in such a way that the press table and the tool are evenly illuminated. However, light should not be directed into the operator's eyes.

For shop lighting both incandescent and fluorescent lamps are being used. Each has its advantages and drawbacks. In difficult cases the best solution may require the advice of a lighting expert.

A correct, systematic maintenance schedule must be devised and scrupulously observed. One of the chief points of maintenance is periodic cleaning of lighting fixtures, walls, ceilings, and other light-reflecting surfaces.

Color

A detail that is intimately related to illumination is the color of walls, ceilings, floors, pillars and columns, machines, and equipment in the shop. As is well known, any surface in a room absorbs a certain quantity of light, depending on the color of its paint. Broadly speaking, the darker the color, the more light absorbed (very dark walls may absorb up to 75%, a light green wall only 30% white even less, only about 15%).

The proper rational, functional application of color in the shop is a science in itself, and is included as one of the basic factors of productivity. It is important not only because of improved illumination, but it also affects production efficiency directly by increasing output, improving product quality, improving employee morale, increasing employee well being, etc. In addition to improving lighting conditions, its specific objectives are safety, identification, and decoration.

The basic principles of correct, scientific application of color schemes for machines and shops may be briefly stated as follows: the essential parts of machines are painted in appropriate colors, which do not clash. Thus, the worker's eye is directed automatically toward his job. The walls behind the machines are painted in eye-resting colors that do not significantly contrast with the colors of the machines, so there is reduced eye strain. (By painting the walls and ceiling in modern light colors—green, orange, etc.—employee fatigue is decreased and production efficiency is increased.)

Referring to safety, it has been found in actual practice that correct distinctive color painting of moving and stationary parts of machines greatly helps to reduce accidents (they create in effect, automatic warning systems). For example, in many stamping shops press bodies are painted in gray or light green and the flywheels are yellow or vice versa, i.e., the body is darker and the flywheel and press ram are painted in lighter colors.

The color of the usual stock stamped in a given press also has a certain influence on the selection of the painting of the equipment. One should choose for the machines and possibly also for the press tools colors that contrast with the work in order to increase its visibility and decrease eye strain.

Colors are often used also for identification of certain parts, controls, containers, pipes (different colors for water, steam, air, chemical substances, etc.).

One of the most interesting and useful advantages of the proper application of functional color is its aesthetic aspect. A correctly painted plant gives the overall impression of order, cleanliness, and good housekeeping. Its consequences are improved morale and increased efficiency (consequently, less absenteeism and less labor turnover). It builds up more interest in the job and in the equipment and creates a desire to maintain the shop and the machines in a clean and orderly condition.

Remember that the machines and attachments must be painted anyway from time to time as protection against corrosion. Since one paint does not cost much more than another of the same type, it is always worthwhile to choose one that gives a correct functional color arrangement.

Color dynamics is often a delicate and to a certain extent, controversial matter and is best handled by experts. Therefore, when in doubt, consult paint manufacturer's representatives; they will be glad to give helpful advice.

Temperature Control

To work at highest efficiency—other conditions being equal—the room temperature must be comfortale, neither too hot nor too cold. For best physical well being, there is no substitute for air conditioning. It means essentially clean, fresh air, controlled temperature, and controlled relative humidity. However, the great majority of industrial plants (especially the existing, somewhat older, and/or comparatively large ones) are not particularly suited for the installation of air-conditioning equipment. In such shops the room temperature must be properly controlled by means of good ventilation in summer and adequate heating in winter, in order to keep the temperatuire around 65–70°F (18–21°C), which is the most comfortable range for performing light manual work.

Windows exposed to direct sunlight can be painted with a special white paint or covered with a special solar-control film. These let through most of the light but keep out up to 75% of solar heat.

Air Contamination

Keep in mind that material-handling equipment driven by internal combustion engines produce exhaust that pollutes the air in the shop. Therefore, give absolute preference to electrical trucks, unless the plant has an excellent ventilation system.

Cooling

Ventilation may be artificial or natural. For artificial ventilation, proper air exhausters or ventilator fans are used. For natural ventilation, in order to avoid disagreements, it is necessary to establish who decides about the opening and shutting of windows, skylights, doors, etc. In most factories, seniority among employees is the governing factor. Seniority usually means an older worker, and it is best that these older workers decide about drafts, because they are more sensitive to them.

As an additional means for cooling the shop in hot weather, it is suggested that flat roofs be painted with aluminum and in addition means may be provided for sprinkling or even flooding the roofs.

In case of air conditioned shops, if the outside temperture is quite high, it is recommended that the inside temperature be kept only about 10 degrees lower than the outside temperature.

Heating

There are several methods of industrial plant heating: central warm air, hot-water radiators, low-pressure steam radiators, steam convectors, high-pressure hot-water convectors, direct-fired space heaters, electric unit heaters. Only an expert is in a position to recommend the most efficent system for providing comfortable working conditions with minimum energy costs.

Whichever heating system is chosen, it is recommended that the temperature control be made fully automatic or at least semiautomatic in order to save labor, power, fuel, and water costs, while at the same time providing a much more uniform temperature.

Noise

By definition noise "is an unwanted, unpleasant loud sound." Noise control is one of those important factors which contributes substantially to improved physical working conditions. In fact, uncontrolled, excessive noise may produce occupational deafness, make people irritable, impair all faculties, and thus lower production efficiency. By proper reduction of noise, efficiency goes up, product quality is improved, and absenteeism and labor turnover drastically reduced.

Noise pollution may be combatted or noise control may be effected in a few different, yet sometimes concurrent and interrelated ways: (a) by reducing it at the source; (b) by confining it around the machine by means of housings or isolated rooms; (c) by dampening its effects away from the source by trapping airborne sound with sound-absorbing materials; (d) by protecting the operators individually (with ear-plugs, or similar devices).

a. Of course, attacking the noise at the source is the ideal solution and makes the most sense. It is usually the most effective and efficient. This means that the presses must be modified accordingly and the press tools must be designed by keeping in mind the special requirements of low noise. Both details belong to the province of specialists and are outside the scope of this volume. In case of necessity it is recommended that consultants on industrial noise control be contacted.

b. The second approach is confining the noise around the machine by creating a sound-isolating curtain. This may be accomplished either by stationary (full or partial) enclosures or by movable shields located between the noise source and the operator's ears. Such enclosures are especially effective in the case of fully automatic presses which work without direct intervention of people.

c. If the first two methods are impractical or uneconomical, the room, at least, must be acoustically treated. Then, the workers tending the presses get very little, if any, relief from immediate noise, however, the general noise level is reduced. This treatment is effected by using on the ceilings, floors, and walls proper acoustical material which absorbs a great part of the noise, without reflecting it. Consequently, reverberation is greatly reduced. Such acoustical materials usually consist of soft fibrous substances (e.g., rubber-coated glass fiber) in the form of panels. These panels are then nailed or cemented to walls and ceilings or sometimes simply suspended.

d. In special cases the operators may receive ear plugs or muffs. These devices considerably reduce exposure to high noise levels.

Floors

Inadequate, uneven floors with ridges, bumps, holes, ruts, etc., decrease productivity because they impede the material-handling (transportation) equipment, cause fatigue to workers, may even cause actual loss or damage to tools and to products, and may cause accidents to employees and power trucks.

An ideal floor for industrial plants should have the following characteristics: high wear resistance; uniform wear, without chipping or cracking; high resistace to chemicals, acids, oil, grease, etc.; strictly waterproof; sound and heat insulating; incombustible; resistant to charging with static electricity; nonslip surface (even after long wear); easy to clean and to service; and low cost.

The usual flooring materials are: cement, steel plates, asphalt, rubber, plastics, etc.

It is recommended that cement floors be painted to reduce dust.

Seating

Provided that the work allows a sitting position for the operator, proper posture chairs (adjustable by the worker), or at least stools of such a height that they permit a natural body position for the operator and thus an easy and free operation of the press, should be provided. Through correct design of the chairs fatigue is not only reduced, but good health is conserved and, consequently productivity is increased.

Regular inspections must ensure correct conditions of repair and maintenance of seating equipment. Workers should be cautioned never to use a chair or a stool as a storage rack, or to put tools, materials, scrap, tote boxes, etc, on it.

Lockers and Washrooms

Often small, insignificant details have far reaching consequences. Therefore, the poor conditions of the shop's sanitary facilities may destroy employee morale. Simple, good housekeeping principles should be applied in washroom maintenance, chiefly cleanliness and perfect working order of every fixture and appliance in it. Sufficient hot water for personal washing should be provided. It saves soap, reduces wash-up time, and helps avoiding dermatitis.

First-Aid Room

A well-equipped, perfectly sanitary first-aid room, where injured employees can be attended properly should be provided. When the plant size warrants it, a registered nurse should be in regular attendance with a doctor on call. If the plant is small, some of the employees or supervisors could be sent to proper training courses for first aid, while forbidding untrained employees to attend each other, except in extreme cases of emergency.

CHAPTER EIGHT

Shop Personnel

Job Specifications

In order to avoid mistakes, omissions, and other kinds of problems—but above all, in order to avoid misunderstandings and, therefore, bad feelings—it is indispensable to spell out exactly in writing what are the duties and responsibilities of the shop personnel. Chart 8-1 gives a detailed job specification for punch press operators. Chapter 9 is actually the job specification for die setters.

Psychological Atmosphere

Agreeable and congenial physical working conditions and surroundings (see Chapter 7) are not enough. Employees are human beings and, therefore, they also are influenced strongly by emotional and sentimental factors. This means that it is of the utmost importance how they are treated by their superiors and colleagues, and what are the general atmosphere and morale in the shop.

The whole mental attitude toward his or her job has often a greater influence on productivity than the introduction of new methods and techniques of production. A satisfied, well-treated worker who enjoys his or her job and is proud of it, may contribute considerably to the improvement of productivity.

Such a proper attitude of the worker is achieved only by suitable efforts of management to establish correct human relationships on all levels. Through alert, sympathetic, and wise handling of people—each one according to his or her own personality—more gains are realized than many executives would believe.

In some plants, management places too much importance on strict supervision, believing that it is a good means to obtain high productivity. In the author's opinion it is true that supervision is a necessary activity; you cannot leave the workers completely alone. However, supervision by itself can never get results without the good will and the positive attitude of the workers.

Today, it is very gratifying to see that there is an overall tendency by industrial management everywhere to improve the worker's environment. Plants which offer the best, most pleasant working conditions, often get the best labor force and are able to keep it. In such conditions the improved morale among the workers results in a tangible increase of productivity.

High production of high-quality products depends on employee morale. The worker is the key to the problem and almost everything depends on his or her attitude toward the work. The morale of the employees is especially important in small

Chart 8-1: Operational Procedure for Press Operators

a. *Before starting a new run*

1. Check the job and if you find any unsafe condition at the press, die, or accessories, do not start working, but inform your supervisor.

2. Before the actual start, always hook on your pull-out guard hooks (if they exist). The same goes for every new start after interruptions (coffee breaks, lunch breaks, visits to the washroom, etc.)

b. *During the run*

3. The basic job for a punch press operator consists of operating a power-driven press. The goal is to cut, form, emboss, punch holes, and make impressions in sheet metal stock. The operation is performed by inserting metal under punch or die and against stops and gages, and starting the machine causing the upper punch or die to descend, performing the required press operation.

4. Follow strictly, in every detail, the instructions received from the die setter and/or supervisor. Periodically check products for dimensional accuracy and quality. If you find something wrong or even suspicious, stop the press and inform your supervisor (for example, excessive burrs, missing holes, deformed part sections, slug pulling, loosening of tool members, excessive lubricant spreading, etc.). Fill containers with the quantities indicated on the production card.

5. Observe your press and die for uniform operation and function. Train your eyes and ears in such a way that you may detect any irregularities that may happen.

6. After an operational accident or mishap you must require inspection of the next piece before resuming regular cperation. Report to your supervisor. Mishaps are miscuts, partial cuts, double blanking, double punching, improper or missing lubrication, etc.

7. In case of any other unscheduled production interruption (lack of stock, lack of lubricant, lack of containers, etc.) immediately inform your supervisor.

c. *After finishing the run*

8. Report immediately to your supervisor.

Note: If for *any* reason you must abandon a press, before leaving it, *shut off* the driving motor. Failure to comply with this safety rule will be considered as a severe offense.

shops where the good will, cooperation, and efforts of each member of the productive team has a marked influence on productivity.

Morale may be influenced notably by physical and emotional environmental conditions. It is not a case of pampering or coddling the workers—as is erroneously assumed sometimes—but of dignifing the work and making it pleasant and attractive by eliminatng or minimizing as much as possible any disagreeable conditions connected with it.

Press Operators' Contribution to Improved Productivity

Perusing books and magazine articles dealing with the production of metal stampings, one comes to the conclusion that the secret of successful stamping production consists simply of correct stamping design and sound die design. The rest is routine. Nothing could be more wrong. It is true that without properly designed stampings and press tools, the production of stampings cannot be satisfactory. However, there is another highly important factor in the manufacturing process of stampings: it is the press-shop personnel, particularly the press operators.

The press operators may contribute—and it is expected of them to do so—considerably to production efficiency. Both direct and indirect, i.e., active and passive, cooperation must be given by them in order to prevent problems and waste and so keep the overall economic efficiency of production as high as possible.

There are several details that are more or less decidedly the operator's responsibility in this respect. The following short list contains the main points (there is no significance to the sequence in the list; every detail is of utmost importance; none of them can be neglected or disregarded): error-free operation; careful stock and product handling; prevention of breakdowns; product inspection; stock lubrication; tool lubrication; tool and press cleaning; and avoiding potential trouble spots.

Operation

Press operators are generally classified as unskilled workers. Therefore, it is assumed that any person with very short, superficial training and preparation may become a press operator, because the work supposedly requires only very little technical knowledge, skill, and experience. This belief is only partially true. In fact, a first-class press operator must know many things, must have adequate experience, and, above all, must be responsible. After all, the shop entrusts him with a tool that costs hundreds (or even thousands) of dollars and a machine that costs thousands of dollars. And it is easy to make a mistake which can cause damage, the repair of which may amount to a very large sum (especially if profit loss, due to downtime, is also included).

Press operators must learn one basic rule: To follow instructions scrupulously, exactly, unquestioningly. They must learn how to use the press controls (first of all how to stop the press!); how to prevent accidents (to persons and to equipment); how to move economically, with the least expenditure of energy, effort, and time; how to handle materials (feed raw materials and remove parts and scrap, etc.); and any other necessary operation which is part of a satisfactory job performance.

Careful Handling and Stampings

One contribution of the press operator to improve productivity consists of reducing rejects. There are several ways to do this, one of which is by taking extra care in the handling of finished stampings.

In case of delicate—weak and/or highly (surface) finished—workpieces, the operator should not throw the stampings into the tote boxes. It is best in such cases to stack the parts on the bench, or on the bolster plate of the press, if there is enough free space, and then place them carefully into the containers. Otherwise, the blanking and/or punching burrs and/or small slugs may cause irreparable damage (scratches on the highly finished surfaces or deformations of the weak portions). Also in the case of delicate stampings it is good practice to line the containers with paper.

In the case of interruptions in the production run, cover the bins containing parts or blanks in order to protect them from dust, dirt, or foreign matter. It is even better, if the interruptions are long, to remove the containers from the work area, in order to have more room for the necessary repairs, inspections, adjustments, etc.

Finally, sloppy handling must be strongly discouraged. For example, it sometimes happens (it should never occur, but it does!) that in containers half filled with stampings, other smaller tote boxes (filled with smaller stampings or scrap) and added. In such cases the original stampings may be damaged not only because of the weight of the smaller tote boxes, but also because they may become contaminated by dirty bottoms of the smaller boxes.

Breakdown Prevention

Active contributions to prevent breakdowns and/or rejects may be summarized as follows:

1. First operation tools. Care must be taken to position the strips or coils correctly (with respect to stops) otherwise too much scrap and incorrect stampings are produced (miscuts, mishits). Double-hit (double blanking; press repeating) may cause serious problems, at least, work spoilage and tool and machine damage.

2. Second operation tools (both cutting and forming). Avoid faulty nesting. The operator must be sure to follow instructions regarding workpiece orientation and burr side. He must be sure to feed the next blank only if the previous stamping is removed from the tool. He must absolutely avoid feeding two blanks together into the die.

3. In case of any—even the slightest—problem or even a suspicion of irregularity (e.g., loosening of tools or tool members, by vibration; any unaccounted-for noise, etc.) the operator must stop the press and report to the supervisor. If some irregularity is detected early, its repercussions are usually slight and not far reaching.

4. In case of miscuts, double hits, or other mishaps, the operator must request inspection of the next piece before resuming production.

5. If the investigation of the mishaps shows that the operator is to blame, then he should be told the extent (in dollars, in toolmaking time, and in press downtime) his mishap amounted to. If there is a profit sharing incentive plan in the plant, then he will be properly impressed. In case of repetitions, proper disciplinary action must be taken.

Inspection

Before starting and during operation, the operator has to look out for possible causes of trouble. First of all, he must carefully inspect the stock for cleanliness in order to avoid rejects. Also any other stock imperfection must be detected before starting to work.

During the production run the operator must inspect the product and the scrap skeleton periodically and in case of defects and/or imperfections, stop the press and report the problem to the supervisor.

The press operator is, of course, not an inspector. The operator is not supposed to check those stamping details that require the use of precision measuring or inspecting instruments, but only such stamping features that can be inspected visually (and/or by touch). Here follow a short list of defects and trouble signs the operator must look out for. And in case of trouble (or only a suspicion of something wrong) he must stop the production and inform the proper authority about what is going on.

a. Stock: Dirty, too oily; scratches; laminations; wrong rolling grain direction; etc.

b. Flat stampings: Dimensional inaccuracies; missing of some holes; incorrect hole locations; "fired" side walls; burrs (height, uniformity); lack of flatness; bruised pilot holes; nicked or chipped cutting edges; slug pulling; slug jamming; scratched surfaces; etc.

c. Second operation (cutting and forming) dies: Additional checking; inaccuracies in the nesting of blanks.

d. Formed parts: Pressure marks; cracks, indentations; tearings, etc.

e. Drawn parts: Dimensional inaccuracies; cracks, ruptures, tearings, burst shells, wrinkles; separations of shell bottom; burnished shell walls; bulged shells; scored shells; uneven shell heights; etc.

Stock Lubrication

In the case of manual lubrication, it is the operator's duty to lubricate the material to be stamped. This has to be done with scrupulous regularity otherwise problems will arise. He should use the right kind of lubricant; the correct application method; and at the right intervals (everything according to instructions received from his supervisor).

Of course, the shop manager should always prefer fully automatic stock lubrication. The consequences of operator's oily and slippery hands are accident hazards and a high probability of product rejects and excessive waste.

Tool Lubrication

This detail is one of those activities which is rather often neglected. Die members that move in guides (small punches in their bushings, die-set leader pins in their bushings, etc.) must be lubricated properly, just like any moving machine member, in order to prevent scoring. Galled or seized tool members require labor, time, and other expenses for reconditioning; but the chief problem is the downtime of the press and of the tool, which means a decrease in production efficiency.

Of course, a fully automatic or semiautomatic lubricating system is the best; if either is impractical (or too expensive), then manual lubrication must be used. For this latter case a well-known (but little practiced!) basic rule must be followed: "Better a few drops often, then abundant streams infrequently." It should be mentioned that oil-base lubricant when employed to coat the stock also lubricates some of the die members as well.

Semiautomatic lubrication of punches is best performed in the following way: suppose that the punch is snugly guided in its bushing in the stripper (otherwise there is no need at all for much lubrication), make at the top of the bushing a small well which is filled at fixed intervals (once or a couple of times per shift). The lubricant will trickle down dropwise to the cutting edges of the punch (Fig. 8-1).

An analogous system is used for the lubrication of leader pins of die sets (commercially available products).

Another, rather simple method of semiautomatic lubrication of punches, consists of putting felt rings around them just above the stripper plate and keeping the felt rings constantly saturated with lubricant (Fig. 8-2).

Fully automatic tool lubrication consists essentially of providing an oil reservoir from which the lubricant is injected to the strategically important points by means of a small pump actuated by the press (ram, crankshaft, or slide) or by some other suitable means. The lubricant may be delivered intermittently or continuously, at an adjustable rate of flow or jet.

In some special cases, the lubricant is fed through the punch body, itself, similarly to what happens sometimes with large twist drills (Fig. 8-3).

For those cases where lubricant is not allowed (e.g. blanking or punching of some nonmetallic materials, such as paper, plastics, etc.), it is recommended that the bushings be made from some adequate antifriction bronze alloy.

Periodic Cleaning of Dies

During the operation with a press tool, small metal particles, slivers, punchings, chips, burrs, dirt, dust, foreign matter, etc., are deposited on the die plate or on the stationary stamping or forming dies. These are potential sources of serious trouble. They are especially apt to damage cutting edges or surface finishes, but they also may bring about more severe consequences. Therefore, it is necessary to clean the die periodically (after a given number of hits). This operation is best performed with low pressure air jets which direct the debris to the scrap bin.

The press also must be cleaned from time to time; of course, not as often as the die, but at least once a day (preferably at the start of a shift), possibly also in the

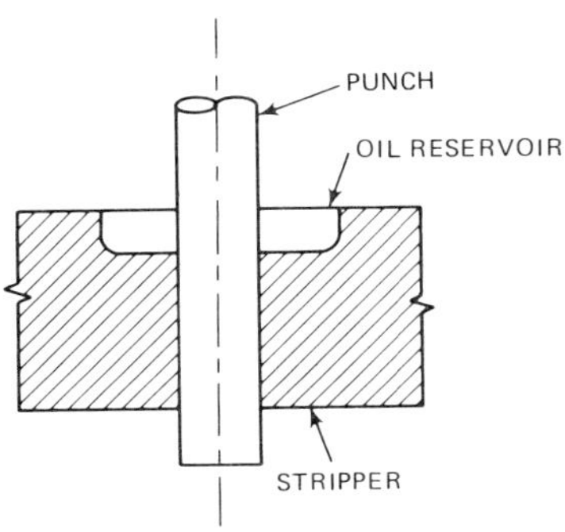

Fig. 8-1. Semiautomatic punch lubrication of long duration.

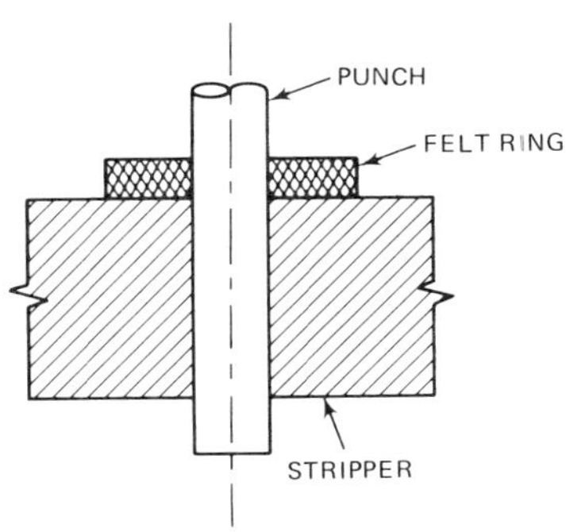

Fig. 8-2. Semiautomatic punch lubrication with felt ring.

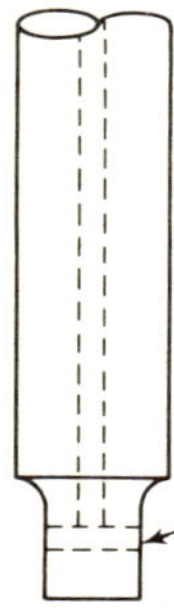

Fig. 8-3. Lubricant fed through punch body.

middle of the shift (before or after the lunch break). The time and effort spent on this activity are repaid manyfold through better press performance.

Avoiding Problem Areas

There are several possible problem areas which decrease the production rate and production efficiency of the press shop. These details are not the direct responsibility of the press operator (they are not his fault; he can do nothing to prevent them). However, whenever they happen, the operator must report them immediately to the supervisor so that corrective measures may be taken.

Here are listed a few of the most common problem areas which the shop personnel must look out for:

Stock is not within easy reach of the operator.

Stock is too oily and so the flat parts sometimes stick together.

Preformed stampings (which must be processed further) are interlocked, due to their shape.

Strips or coils are of incorrect width or are deformed so that they stick between the stock gages.

Feeding or unloading devices inappropriate for the job and/or not in satisfactory working condition.

Blanks have too much burr so that introduction into the die (nest) is difficult and/or unsafe.

Press or tool actuating mechanism too hard or unsafe to handle.

Wrong lubricant used.

Stock of unsatisfactory quality.

Unreliable functioning of some safety device.

Continuing Education

There is a saying to the effect that "He who stands still, remains behind." In fact, with the enormous speed of technological progress, it is absolutely necessary to keep up one's professional training and preparation. For this purpose it is good practice to organize from time to time short refresher courses for die setters, diemakers, inspectors, and other higher graded personnel.

For tool engineers and others who want to improve their skills, many progressive companies have established the following policy: The tuition fees are paid by the company for previously approved technical institute, college, and university courses, after their successful completion. This also holds for correspondence schools, programmed learning, and similar scholastic activities.

A practical suggestion: Every industrial company receives many free trade and professional magazines every month or week. They contain excellent up-dated information about engineering and technical subjects. Their content is very useful for the continuing education of technical personnel. After management and engineering personnel are through with these magazines, they are put on a table near the employee exit door. Above the table, on the wall, there is a large poster saying: "Free magazines—Help yourself—No obligation to return" or something to that effect. A responsible office employee takes care of always keeping the table well stocked.

Working Fatigue in the Metal Stamping Shop

Alert and conscientious shop managers should be concerned with every detail that may affect employee welfare and production efficiency. A somewhat neglected detail in this respect is the study of fatigue, its causes and its remedies. The time and effort spent in keeping fatigue to a minimum are well worthwhile in terms of production-efficiency improvement.

In order for punch press operators to reach their highest efficiency, it is necessary to do everything possible to prevent premature fatigue or to eliminate it when it occurs.

Fatigue is a rather difficult term to define, because it refers to a subjective sensation which it is not possible for an observer to perceive directly or to measure and to express numerically. Its presence is made evident through its symptoms.

The most characteristic and most common symptoms of fatigue, easily recognized by the woker are: overall bad feeling; reduced desire to be active, reduced muscular force, muscular ache (general or localized in a given part of the body), decreased capacity of concentration and of attention, gastric problems, and irritability.

Obviously, the rate of intensity of fatigue varies within quite wide limits betweem different individuals. The fatigue rate is affected by objective factors (type and details of work and physical working conditions, such as temperature, air humidity, ventilation, lighting, vibrations, noise) and by subjective factors (age, sex, physical constitution and condition, training, experience, hunger, exposure to certain chemical and physical agents).

Since the consequences of fatigue are both a decrease of production efficiency and an increased proneness to accidents, it is imperative to study thoroughly the causes of fatigue in order to know which are the corresponding remedies.

The causes of fatigue may be divided into two groups: material and psychological.

Material Causes of Fatigue and Remedies

Excessive and inappropriate physical exertion for the execution of a job is one of the chief causes of fatigue: remedy—work simplification. (See corresponding item in Chapter 6.)

Moreover, practical tests and actual experience have demonstrated that to delay the onset of fatigue, in addition to reducing physical activity, one of the best means consists of ensuring correct blood circulation. (With a thorough, unhindered circulation, blood is carried to every working muscle in the body and thus can supply the necessary replacement energy.) It is interesting to note that static fatigue is often more intensive and sometimes occurs more quickly than dynamic fatigue. This is because immobilized, contracted, and tensed muscles become tired more easily. Consequently, in certain cases, it has been found useful to modify the work routine, even introducing additional (not strictly needed) movements to the work cycle, just to avoid static fatigue.

Another, quite easily and commonly practiced method for activating good blood circulation is to allow both sitting and standing while working. In fact, both positions *alone* lead to early fatigue. So, when the operator feels tired, he may change his position from standing to sitting or vice versa, thus relieving fatigue for awhile. Correct posture chairs must be adjustable in height so that the height of the arms (elbows) be the same in both positions, sitting or standing.

The actuation of machines by foot (treadle pedals) is much easier and less fatiguing in a sitting position. In addition, the hands are not busy with the controls and thus may be used to perform actual work.

The physical environmental conditions (good housekeeping, light, heat control, noise control, etc.) are of vital importance. It has been found in the majority of cases that, if the work is easier and is performed in agreeable comfortable conditions, the worker is less subject to fatigue. His efficiency increases automatically without any increase in tiredness.

A logical consequence of tiredness is the decrease of production efficiency, especially in the later hours of the of the shift. For example, practical trials have shown that in the case of long working shifts (8–10 hours without interruption) the production efficiency in the last hours is less than 50% of that of the first hours of the same shift. Consequently, almost every shop has introduced the so-called "coffee break" practice which often gives a 15–20%, or even more, overall increase in production efficiency.

The "coffee break" consists essentially of anticipating fatigue by interrupting the work every couple of hours (the usual method is one break in mid-morning and one in mid-afternoon; in some shops, even every hour or so) just before fatigue would reach its peak; the interruption lasts a few minutes (it is usually proportional to the intervals between breaks: 5–20 minutes). These pauses may be occupied with taking some nourishment (coffee, soft drinks, snacks) listening to music, reading, smoking, and relaxing.

During these pauses not only is the physical condition of the workers restored, but also their mental and emotional condition are improved. Consequently, their production efficiency is increased considerably, in spite of the "lost" time.

Psychological Causes and Remedies

Every smart foreman and plant manager knows that a person who likes his job, is interested in it, and is happy with it, tires much less quickly than a worker who does not like what he is doing. Therefore, the first, and very important, task for management consists of assigning every worker to that kind of work which agrees with and best fits his temperament, his likings, and his capabilities.

Also personal matters have strong influence on an operator's state of mind, emotions, and well being. The effects of special domestic situations, family problems, or relations with fellow employees should not be overlooked in this context.

Independently from the interest that the worker may have in his job and from his personal problems, there exists always the serious problem of boredom, which is unavoidable in such highly repetitive jobs as the operation of a press. Often, with some types of individuals, the monotonous working at the press or at an automatic machine tool has a hypnotic effect. Exhaustive investigations have demonstrated a few basic facts about boredom, which should be rather obvious. First of all, boredom lowers the production rate. In addition, unskilled work, i.e., jobs which do not occupy the worker's mind, produce more fatigue by monotony than skilled work. Finally, the more intelligent and better educated persons are more easily bored.

Remedies for work monotony are rather difficult to create. One of the little practiced, although quite effective, methods for counteracting boredom consists of freeing the press operator from the compulsion of a constant machine rhythm. Since the worker's capacity of attention and of concentration vary with the fatigue rate, it is advisable to provide the presses with variable speeds. Thus the job rhythm may be adjusted to the momentary physical, emotional, and intellectual conditions of the operator.

In addition, variable speeds have another, important psychological effect on the operator. If he knows that he is not a slave of a machine running at an inexorable rate or speed, then he feels much better about the matter and does not become bored and tired too quickly (sometimes this knowledge alone is sufficient).

Another remedy against boredom consists of job rotation. This means antispecialization, i.e., the workers are trained for more than one job and they are shifted around periodically from one job to another.

Mini-Zero-Defect Program

The "zero-defect" (ZD) philosophy, since its conception and creation has spread quickly among metalworking and other kinds of industrial plants in the United States. Some time ago, when the author was working in Chicago, his company wanted to adopt it in order to improve production efficiency. Therefore, all the available information was gathered from the leading companies which applied ZD, starting with Martin Co., in Orlando, Florida, where ZD was originally conceived, and including the available literature.

In such a program workers are indoctrinated and persuaded by such means as posters and speeches to change their attitude toward their work: striving for total avoidance of mistakes is the ultimate goal. It is a vital sensible, reasonable work philosophy which has given remarkable results, especially in the defense and space industries, where in many cases their products *must* work right the first time, because there is no second chance. The best results have been achieved in large plants. No reports were available about ZD programs from plants having less than 500 employees.

In view of the difficulties and problems—and costs—related to the realization of a complete ZD program in a small plant like the one where the author was working (with only about 100 employees), it was decided to adopt a "mini-program," which demands a small amount of effort and very little expense. The results were quite satisfactory, if not as spectacular as a proper or full-fledged ZD program.

The basic idea consists of rhetorical question and answer notes. It started simply. On a certain payday, every employee received, attached to his or her paycheck, a slip with this text:

> Do you think that mistakes are unavoidable?
>
> If you do, then please think over these questions:
>
> How many times did you accept mistakes in your paycheck?
>
> How many times lately did you accept being shortchanged at the supermarket?
>
> How many times did you leave your car motor running in the parking lot or the driveway?
>
> How many times did you go home to the wrong house?
>
> How many mornings did you put on one brown and one black shoe?

The next paycheck had a new note with this text:

> Your answer to all of last paycheck's questions was probably this:
>
> I never made any of those mistakes!!
>
> Therefore, it may be assumed that errors are related to the importance a person places on specific things. Then, why don't you transfer your personal standards to the work you do, so we will do our jobs with error-free performance?

Starting with the third paycheck, a new reminder is attached. Chart 8-2 gives the text of representative notes.

Chart 8-2: Mini-Zero-Defect Program Slogans for Paycheck Envelope Inserts

Week 1

Do you believe that mistakes are unavoidable?
If you do, then please think over these questions:
How many times did you accept mistakes in your paycheck?
How many times lately did you accept being shortchanged in the supermarket?
How many times did you leave your car motor running in the parking lot?
How many times did you put on one brown and one black shoe?
How many times did you go home to the wrong house?

Chart 8-2 (Continued)

Week 2

Your answer to all the questions of last week was probably this:

I never made any of those mistakes!!

Therefore, it may be assumed that errors are related to the importance a person places on specific things. Then, why don't you transfer your personal standards to the work you do, so you will do your jobs with error-free performance?

Week 3

When you drop a coin, a glove, any object, you always pick it up, don't you?

Then why don't you pick up a stamping dropped on the floor?

You don't leave the water on to overflow in your bathroom, do you?

Then why don't you stop the motor and shut of the air and the lubricant when you leave your press?

Week 4

Supposing you need an operation. Would you want a doctor with a reputation for being almost right?

You wouldn't!!

Then why don't you apply the same standard to yourself on your job?

Why don't you try to do a PERFECT job every day?

Week 5

When you drive on the street, highway, or expressway, you are CONSTANTLY careful because you know that a mistake or distraction may cause damage and injury.

Then why don't you pay the same kind of *constant attention* to your job? Why don't you try hard to avoid mistakes?

Week 6

Misunderstandings can cause a lot of damage, spoilage, waste, and accidents to people and equipment. If you are not sure how to perform a job, *don't do it,* but ask for clear, understandable instructions.

Do it right the first time!!

Make this slogan your way of life. Perform every job as if there were no second chance and no possibility for correcting errors.

Week 7

Are you interested in *job security?* For yourself, your buddies, and others?

Then improve your job performance. In this way, your company's overall efficiency and reputation will be improved. There will be higher demand for your company's products. Customers will be kept happy. Therefore, you will contribute positively to ensure a good permanent job with your company.

Week 8

When a visitor to your home praises its nice appearance and orderliness, you are proud, aren't you?

Then, why don't you do everything in order to be proud of your *housekeeping* of

Chart 8-2 (Continued)

your working place? Everyone knows that an industrial plant cannot be kept clean and orderly without the constant, full cooperation of *every* employee.

Week 9

No job is unimportant!

Sweeping the floor and moving materials are important functions just like operating a press or setting up a die. These latter jobs are more difficult, they require apprenticeships, they involve more responsibility; therefore, they are paid higher salaries.

Week 10

Take part in the *suggestion* activity. Your contributions are always very valuable to your company.

By presenting practical, useful suggestions, you will get satisfaction, recognition, and awards.

This program lasted a total of ten weeks. The results were tangible and positive: a decided improvement in overall efficiency, avoidance of mistakes, and better housekeeping. It was everybody's opinion that the experiment was a complete success and well worthwhile.

Referring specifically to employee morale, what was gathered through the usual "grapevine" was this. The people liked the challenging tone and personal appeal of the notes. These were more direct and more immediate than slogans on posters scattered around the shop. The notes gave them the impression that management was treating them as individuals and not merely as pieces of equipment. Several times the author witnessed small groups discussing a note and when they found out that the author was the originator of these notes, many employees came to him either expressing their agreement or arguing with him about the difficulties of realizing the improvements they felt necessary, but which demanded more help and cooperation from management. On the whole, these notes had a stimulating effect, and this alone should be considered as a positive result regarding employee morale.

CHAPTER NINE

Die Setting

Introduction

Everyone connected with metal stamping manufacturing is well aware of the importance of the setting of dies in presses for the production of high-quality stampings. One can have the most ingeniously designed die, built with perfect craftsmanship, but it will function poorly and inadequately if it is handled incorrectly by the die setter.

In too many press shops—especially in small and medium-sized plants—die setting is not given the importance it deserves. As a basic rule, only qualified (trained and experienced) die setters or closely supervised die setter trainees should be allowed to set up dies, to adjust them, and to remove them from the presses.

Die Setting Activity

The procedure of correct die setting has a definite, logical sequence. If the corresponding steps are followed scrupulously and the pertinent rules observed strictly, in the given order, the results achieved will always be highly satisfactory.

The die setting activity is divided into four basic phases or stages: (1) preparation; (2) setting proper; (3) releasing the setup for production; (4) die removal.

Setting of Cutting Dies

Here follows the exact, chronological list of steps for the setting of a standard cutting die. Simple blanking, punching, progressive, or compound dies require more or less the same basic steps in setting.

Preparatory Activity

1. Provide sufficient quantity of raw material for a test. Of course, the stock is supposed to be checked for proper quality, dimensions, finish, etc.

2. Have the die ready for mounting in the press, prior to or as the press to be used becomes available. Remember that any unnecesary delay or idle time of the press means productivity decrease.

3. Transport the die to the press. Use adequate die-handling equipment.
4. Check the tool visually for correct working conditions:

a. the cutting edges of both the punches and die plates must be correctly sharpened; in addition, no nicks, cracks, chipped edges, nor other damage should be present;

b. quantity and type of punches;

c. slug and blank holes in die plate are not obstructed;

d. auxiliary mechanisms, e.g., ejectors, movable strippers, stops, pressure pins, locators, stock gages, guides, springs, bolts, etc., are in working order;

e. free movement of leader pins and bushings of die set (no scored or galled pins);

f. loose or missing screws;

g. undamaged die shank.

5. Check the characteristics of the specified press for size (sufficient tonnage capacity to supply a cutting force of 50–100% more than that required for the job is considered good practice); speed; stroke length; shut height; die space; attachments (feeding and ejecting devices; cushions; guards; etc.).

Setting Proper

6. When the press is free (the previously used die has been removed and carried away), the first step consists of taking the necessary safety precautions for the die setter and helpers.

a. Clear the press area of all material (skids, containers, accessories, debris, etc.) to about 3–4 feet all around.

b. Next, shut off the lockout power in the press (use a padlock on the main switch, if possible) while the die is being set up or adjustments are being made on the tool or the press.

c. Shut off air.

d. If possible, keep safety blocks under the ram while working on the press.

e. In many plants, as an additional safety precaution, die setters put a large portable sign or poster on the press, which reads in red: "MEN WORKING ON THIS MACHINE" or "DANGER—DON'T CONNECT" or something to that effect. The power lockout is, however, the best practice. Switch on power again only after all the necessary work has been completed that can be performed without power.

7. Clear the press table of all material (tools, parts, scrap, accessories, etc.).

Check the mechanical working conditions of the vital parts of the entire press while in a standstill condition; never when operating. Inspect carefully for loose, broken, damaged, or worn parts. Take special care to check the clutch and the foot treadle clearance and treadle guards. Nothing should be left near the treadle which could cause a problem or accident to the press operator.

Check the tension of the clutch and brake, antirepeat device, play of ram in its way, counterbalance, KO bar, etc. Adjust them, if necessary.

8. After having scraped away dried dirt, if necessary, clean carefully and thoroughly by wiping away with a *clean* rag any oil, grease, dirt, litter, slugs, chips, scale, foreign matter, etc., the clamping surfaces of the press table, as well as the bolster plate, including T-slots and slide (inward, outward, and bottom faces). If necessary, use a solvent to get everything clean.

Clean out tapped holes for fastening in the table and bolster plate. If the press is provided with threaded plugs for these holes, when not in use, be sure that they do not protrude above the table or bolster surface.

Remove with a superfine file, any nicks, burrs, dents, or raised spots on contacting surfaces of ram, table, and bolster plate, and around tapped holes.

9. If not provided with an automatic centralized lubrication device, the press must be lubricated carefully. Always lubricate the machine at standstill, never when rotating. Lubricate (oil or grease) every bearing, guide, gib, etc. Proceed systematically, following exactly the fixed procedure of the corresponding check list, in order not to forget anything (main bearings, slide ways, pitman, clutch, attachments, mechanisms, etc.). Be careful to prevent any dirt entering the oil cups, oil holes, etc. or dirt being carried inward from the pressure fittings. Never employ the brush used for lubricating the stock to lubricate the press or guiding surfaces.

10. Now, raise or lower ram adjustment (shut height) if necessary, to clear the die by about ⅛ inch minimum.

11. The same rule for cleanliness and smoothness of contacting surfaces also applies for the die as for the press. The bottom surface of the die holder, the top surface of the punch holder, and the shank must be thoroughly scraped and cleaned before setting up in a press.

12. In case of dies having short shut heights, select a suitable bolster plate and mount it on the press table. Be sure that both top and bottom surfaces are clean before mounting on the press table. If there are any raised spots on either the bolster plate or the press table, remove them. Here follow a few basic requirements for the selection of a correct bolster plate:

a. adequate thickness;

b. sufficient openings for blanks and slugs to pass through, at the same time the openings should not be too large; there must be enough support for the die;

c. properly located tapped holes for the fastening of the stationary and movable parts of the tool; it is good practice to plug the temporarily unoccupied tapped holes in the bolster plate and ram with threaded plugs or at least with cloth or waste, or wooden plugs, in order to keep them clean.

13. Be sure that the air cushion pins are level with the bolster plate top.

14. If air cushion is not used for the die in question, cover it with a metal sheet in order to keep dirt from entering.

15. Instead of or in addition to bolster plates, simple parallels frequently are put under the die shoe to raise the tool to the required height. These parallels must have ample width and should be located as near as possible to the clearance holes of the blanking and punching openings (without, of course, obstructing them even partially), in order to give the greatest support to the die and to keep the die shoe from sagging.

16. Lift the die to the press table or bolster level, either manually or mechanically. Slide the die on the press bed and locate it—manually or with a pry bar—

without fastening, correctly with respect to the ram and other devices (feeding, etc.).

17. Check the alignment of all clearance holes for blanks and/or slugs through the die shoe, bolster, and press table. For this task the best aids are a flashlight and sometimes also a mirror. In the case of large dies, the blanks and slugs are usually removed either by simple gravity (inclined press) or by means of air jets. In such cases the bolster plates are usually "blind," i.e., there are no clearance holes for large stampings and punchings.

18. During the handling of the die, the punch holder should rest on proper bumper blocks, spacer blocks, bumper pins, etc. (or, in an emergency, on standard parallels). These accessories prevent the accidental dropping of the punch holder, thus injuring the die setter or other helpers working on the press and endangering delicate tools (damaging stamps or frail punches or other tool members, ruining a stationary stripper, causing problems by having a cutting punch enter the corresponding die opening too much, etc.).

These spacing blocks must be even in number—generally two are sufficient—and symmetrically placed. They are made either from mild steel, or, even better, from some light metal such as an aluminum or magnesium alloy. Their cross section may be simply rectangular or an extruded structural shape. Sometimes they are also made in the form of rings to be put on the leader pins of the die sets. They are usually commercially available although sometimes they are "home-made."

If no spacer blocks are available and the die has a stamping or a delicate forming station, place the correct thickness of metal sheet along said stations, in order to prevent die damage during setting up.

19. Turn on power for a short while, just enough to turn over the flywheel once or twice.

20. Shut off power.

21. By taking advantage of the flywheel still turning slowly due to inertia, inch the slide down until the ram bottom contacts the punch holder top. Be sure that the ram rests firmly on the punch holder.

22. Adjust the shut height approximately, leaving it slightly *higher* (about 1/16 inch) than actually needed.

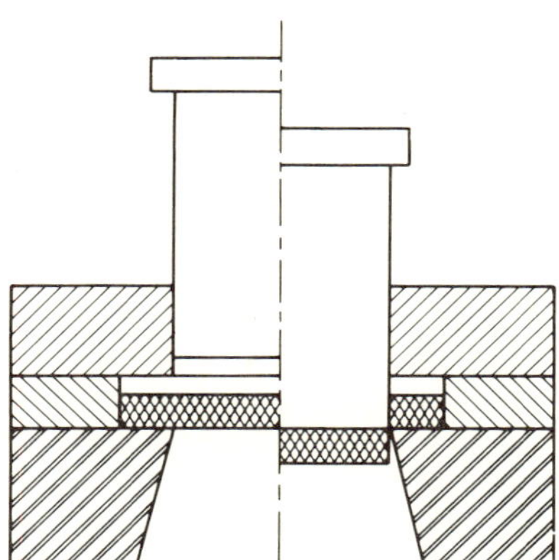

Fig. 9-1. Correct adjustment of ram stroke means minimum punch penetration in die opening at the stroke bottom and minimum separation of the punch from the stripper bottom.

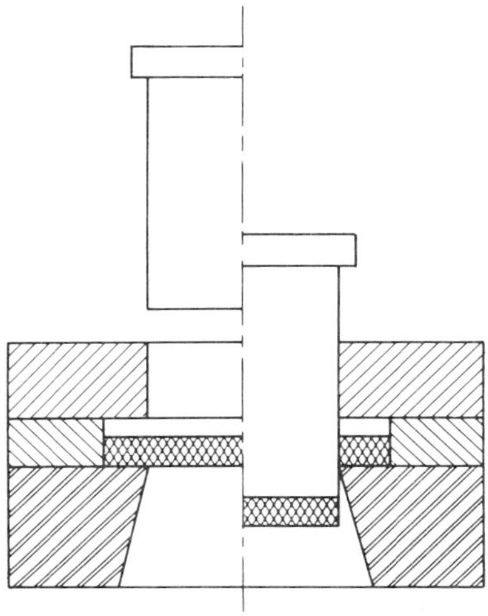

Fig. 9-2. Too long ram stroke produces exaggerated punch penetration in the die plate with the punch leaving the stripper. This causes several types of problems.

23. Securely fasten the top part of die to the slide. Small and medium-sized dies have shanks which are secured in the corresponding hole in the ram or in a jaw fastened with two bolts. In the case of large dies where there is no shank, the punch holder is fastened to the ram indirectly by means of clamps or directly by means of bolts engaged in tapped, spotted holes in the ram bottom. The length of thread engagement of these bolts should be at least 1½ times the bolt diameter.

24. Securely clamp the stationary part of the die to the bolster plate or press table.

25. If the *stroke length* is adjustable, then make it the shortest possible, commensurate with the working requirements of the die. An unnecessarily long stroke means increased wear of the press and of the tool, unfavorable pressure transmission conditions, increased use of lubricants, increased accident hazards, etc. Figures 9-1 and 9-2 show correct and incorrect stroke adjustments, respectively.

26. Referring to the *position* of the ram stroke, the basic conditions for correct adjustment are:

a. The lowest position of the stroke corresponds to the shut height of the die. This means that the blanks and slugs must be safely separated from the stock. However, the shortest punches should not enter into the die openings more than is strictly necessary (about 0.010–0.020 inch). Whenever possible, try to avoid metal-to-metal contact between punch and die plate.

In the case of progressive dies with a forming or stamping station, the lowest position, of the ram must be just enough to perform the forming operation while avoiding hitting too hard.

b. At the highest position, the stock must be separated from the punch(es). The longest punches must stay at least 0.100 inch above the stripping surface, that is, the bottom surface of the stripper plate. Be sure that the punch holder never touches the press gibs.

c. The pitman screw should be unscrewed very little, possibly only a few threads, but under no circumstances more than 50% of the threaded length. If the shut height of the die is too low, use adequate bolster plates and/or riser blocks or simple parallels.

d. If there are bumper blocks or bumper pins assembled to the die, these should not be hit hard, possibly not even touched. Check with small pieces of soft solder; these should be at least 0.006–0.010 inch thick after a hit.

27. To adjust the position of the ram stroke, the locking nut of the pitman screw must be loosened first. After each adjustment, the nut must be tightened securely before the next steps. Failure to comply with this requirement may result in consequences that are sometimes devastating: overloading of the pitman screw threads, overloading of the die and the press crank, and even breaking the tool and/or the press itself.

28. Connect and adjust air pressure if necessary.

29. Raise the ram and lubricate all the moving parts of the tool: grease the leader pins, oil cutting edges of the punch and die plate, small guided punches, and oil slides, if there are any.

30. In the case of small presses, turn the flywheel slowly by hand a few times in order to be sure that everything is all right. Put on the belt (if it has been removed at the start of the die setting; an operation many die setters do with small and medium-sized presses).

31. Start the driving motor, after having connected the power, and let the ram ride up and down several times, without cutting.

32. In the case of a press with adjustable speed, select and adjust the proper speed value.

33. In the case of first operation dies, stop the motor and position the strip or coil stock for hand or automatic feeding. Pass the stock through straightener rolls, feeding devices, and decoiler.

34. Check ease of feeding by passing the stock between the die-set leader pins and the clamping straps and bolts or nuts. It must pass freely and easily. Check the whole path, i.e., including the strip skeleton, the exit side of the die.

35. In the case of second operation dies, the part-locating nest must allow easy, trouble-free introduction and removal of the parts. If the nest is too tight, the production is slowed down and frequent misfeeding may occur.

36. Check the correct, safe functioning of ejectors, knock-outs, strippers, etc. If they fail, the consequences are disastrous for tools and presses.

37. Lubricate stock manually.

38. Make a few trial parts. Check and inspect them visually, with care, for any and all irregularities, such as dimensional inaccuracies, excessive burrs, missing holes, slug return, etc.; study shear marks and breaks around the blank and hole peripheries. The tendency to burr formation and the appearance of the sheared surfaces give a good indication of the die setter's competence and craftsmanship.

39. Make the necessary adjustments (e.g., correct cushion pressure and proper position of KO bar) until parts are satisfactory to you.

40. Submit several sample parts to Inspection for final check of first parts. In some plants the diemaker does not wait for Inspection to pass the parts, but leaves them with Inspection and returns to the press to finish the setting job. In other plants, it is preferred that the die setter observe the inspecting operation and so learn

more about this activity. (This latter practice is more practical also from the job efficiency standpoint by avoiding of eventual repetitive work in the die setting.)

41. After the parts have passed inspection, finish the die setting job, performing operations 42–51.

42. Check whether power is shut off.

43. Retighten every screw, bolt, and nut, including the pitman lock-nut, fastening bolster plate to press table, etc.

44. Recheck alignment of punch and other die members, since, while retightening the straps, the stationary part of the die may have been shifted.

45. Activate the automatic or semiautomatic feeding and unloading mechanisms and devices (chutes, slides, tables, magazines, stacking attachments, bins, tote boxes, separate strippers, air controls, conveyors, blow-offs, etc.) and automatic stock lubrication devices.

46. Check, especially, and adjust automatic-feeding devices for correct feeding lengths (center distances).

47. Replace and adjust all safety devices (guards, leader pin protectors, limit switches, etc.). These should provide maximum protection for the operator, with minimum interference of free operation. Check the accident prevention devices, in order to see whether they are in perfect working condition.

48. Run a small quantity of parts (100–500 according to the characteristics of the job) under normal operating conditions, in order to be sure that everything is working properly.

49. Take care of housekeeping (see Chapter 7).

50. Arrange the placement of stools, tote boxes, etc., for best working conditions.

51. Notify foreman that press is ready to run (or to be checked, if this is standard practice in the shop). Inform the foreman about any, even if insignificant, deviation from standard procedure.

Before Releasing a Setup for Production

52. Recheck everything: die set, press brake, antirepeat devices, safety devices, knockout bar for clearance at top of stroke.

53. Set counting devices to "0."

54. Provide reasonable quantity (at least 2 to 3 hours supply) of stock, lubricant for stock, and containers for parts and for scrap.

55. Check gravity chutes for handling stampings and/or scrap.

56. Check for possible causes of problems:

a. unsatisfactory stock quality

b. distorted, or otherwise defective, stock and parts

c. stock that is too oily

d. stock not within easy reach of the operator.

e. press and tool actuating mechanisms not easy to operate.

57. Before actually handing over the press to the operator, be sure that he understands thoroughly the correct operation of the press and of the die; how the job is to be done, pointing out special problems and difficulties that may arise during operation, safety precautions, inspection, etc.

58. Who instructs the press operator is a controversial subject: in some shops it is

the foreman, in others it is an experienced press operator, and in others it is the die setter. It stands to reason that the last solution is the most reasonable and the one that makes most sense. Who knows more about the setup than the man who did it? Of course, this presupposes that the die setter has the capacity to communicate readily with the press workers.

Die Removal

After the completion of a run and/or in the case of die trouble, the tool must be removed from the press. Logically, the removal operation sequence is practically the reverse of the setting up of dies.

58. In many shops, before removing a die from a press, it is inspected by an inspector, supervisor, or toolmaker, in order to establish whether the tool must be serviced (repaired, sharpened, or reconditioned) before storing. In other shops, it is the die setter who must perform this inspection. In still other shops, the die is inspected after removal, in the storage room before storage.

59. In progressive dies having a stamping or a forming station, place a piece of stock or a part along these stations.

60. Inch the ram to stroke bottom.

61. Remove safety guards and air cnnections. Shut off air.

62. Remove straps from stationary part of die.

63. Loosen shank clamping bolts, or remove clamps holding movable part of die set.

64. Remove chutes and other material-handling aids.

65. Inch the ram to top of stroke.

66. Shut off power.

67. Slide loose die out to lift truck.

68. Clean die thoroughly, removing oil, grease, dirt, etc.

69. Visually check and inspect die again.

70. The last parts and last piece of strip skeleton always go to storage (through Inspection, if necessary) together with the die.

71. Lower die to skid.

72. Clean press (housekeeping)

73. Remove bolster plate, if necessary.

74. Transport die to designated area of storage room, if servicing is not needed. Do not put die in rack, but on a bench or table provided for the receiving, or on a skid on the floor. Only the die storage personnel should place the die in its proper location.

Setting Up Forming Dies

Of course, the majority of the basic concepts and suggestions as outlined for cutting dies, also apply to any other type of die, e.g., forming dies. However, there are a few special details that must be taken into consideration in the case of bending or forming die setting.

1. In the case of open (not mounted in die sets) bending dies, many die setters use a previously bent stamping (produced in the last production run and kept with the bending die in the die storage room) to facilitate the setting of the tool and the ad-

justments of the ram stroke. However, such a technique is not recommended, because, in the case of unsymmetrical stampings, the press ram may be pushed to one side if there is too much play in the slide. It is much better to align the stationary stamp, after having fastened the movable stamp to the ram, with the aid of feeler thickness gages or simply with small metal strips of correct thickness. The same procedure also applies to the setting up of drawing dies.

2. In constrast to that for cutting dies, the ram stroke should be quite long, both in order to ensure ease, speed, and safety of operation (loading and unloading) and to avoid the crankshaft working in the upper sector of the slide stroke. The adjustment of the slide stroke position is made by successive practice trials (beginning with a *too high* adjustment) until the correct position is obtained. The instant when the die is completely closed (with the stamping in its place) must coincide perfectly with the bottom dead end position of the press stroke (shut height). To make these adjustments correctly is a real test of the die setter's skill and training.

Note: If the lowest stroke position is too high, the part will be formed to a smaller angle than specified; while in the case of too deep setting the press will be overloaded, and this may easily damage the tool, the press, and the stampings.

3. Remember that the formed stampings are different when the press is moved slowly by hand or inched by coasting than when the press runs by power at its standard speed. (The impact is different, and, therefore, the bending angles may vary.)

4. Check whether the blanks have too much burr, so that the introduction into the die nest is difficult. Also, preformed stampings (which must be processed further) are often bound together because of their shape. In both cases the production rate decreases; such conditions must be reported to the foreman so that they can be eliminated immediately. The best solution for this problem is stacking the stampings in the previous operation.

5. Check stock thickness scrupulously. Forming and drawing dies are designed and built for specific stock thickness values; tools do not perform satisfactorily when stock thicknesses are too different from those specified. Too heavy stock is especially dangerous causing possible damage to dies or even to preses.

6. If the die is not mounted in a die set and the bending load is not distributed evenly (see Fig. 9-3), use fastening clamps and/or keys or pins, accordingly, to resist any side thrusting force components.

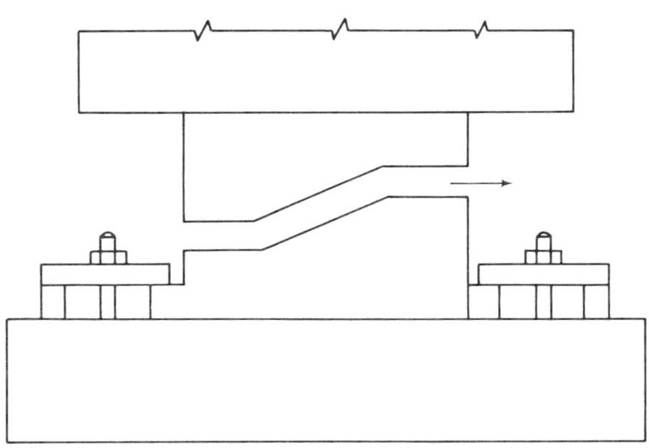

Fig. 9-3. Correct forming load distribution should be strived for in nonsymmetrical stamping dies.

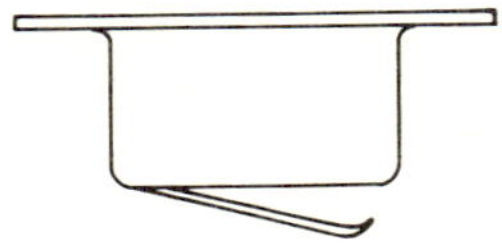

Fig. 9-4. Tearing of drawn shell bottom indicates too high blankholding pressure.

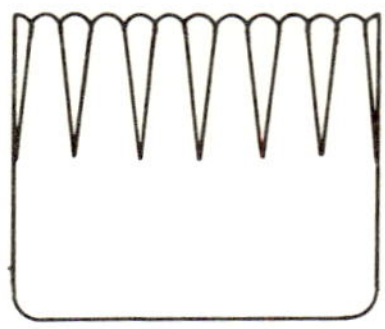

Fig. 9-5. Wrinkles in side walls of shells indicates too low blankholding pressure.

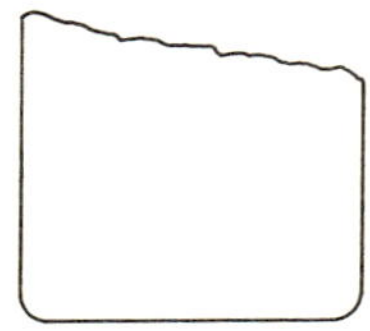

Fig. 9-6. Uneven shell height indicates blankholding pressure is not uniformly distributed.

In addition, if possible, the direction of side thrust should go against the solid guide side of the press ram and not against the adjustable side.

Setting Up of Drawing Dies

In addition to the instruction given for cutting and forming dies, when setting up drawing tools, take into account the following details:

1. The preliminary inspection of the tool must be the most careful possible. Perfect parts cannot be made if the draw die has even the slightest surface irregularities, such as dents, scratches or marks caused by bumping against heavy objects. These imperfections must be eliminated, especially from the active surfaces of the drawing rings, the blankholders, and the punches. If these are not eliminated, the drawn shells will not be satisfactory and the die will have to be removed for servicing.

2. Take special care of the alignment of the punch with the draw ring. Analogously to item 1 of the previous section "Setting Up of Forming Dies," do not use a dummy shell for aligning, but use feeler gages or small strips of sheet metal.

3. In the case of drawing dies, mount the blankholder equipment and adjust it carefully and correctly. Take care not only to avoid too high or too low pressure (with their troublesome consequences) but also, where required (see below), ensure uniformity of pressure distribution on each point of the controlled surfaces (to do this use strips of stock of the correct thickness).

Start with the *highest* (air cushion) blankholding pressure and decrease the pressure gradually until wrinkle formation starts. Then increase the pressure slightly.

Uniformity of blankholding pressure distribution is required only for symmetrical, regular shells. For irregularly shaped shells, the pressure must be higher at some points than at others. Proper preparation of such kinds of jobs is the mark of an outstanding, experienced die setter.

4. Provide the right type of lubricant for the job in question, taking into account

all the pertinent conditions (stock quality, shell shape, size reduction, etc.). In drawing operations the role of the lubricant quality is very important; the correct selection of lubricant requires much experience and specific information. Whenever possible, always consult a reliable supplier of lubricants.

5. Reading a shell. The chief defects with drawn shells are: (a) torn bottom without completing the shell (Fig. 9-4), which means too high blankholding pressure; (b) wrinkles (Fig. 9-5), which mean too low blankholding pressure; and (c) uneven shell height (Fig. 9-6), which means nonuniform blankholding pressure distribution and/or eccentric blank location.

CHAPTER TEN

Feeding in Press Operations

The feeding and ejecting methods of stock and stampings have a decided influence on production efficiency because they practically determine the production rate, i.e., number of stampings per hour, day, shift.

In metal stamping manufacturing, we distinguish the following basic production methods:

1. Fully manual (both feeding and removal performed by the operator).
2. Semiautomatic (the intervention of the operator consists of starting each operation; the rest of the operation cycle is performed automatically by the press).
3. Fully automatic (both feeding and ejection are performed by the press and/or attachments; one operational cycle follows the other continuously, without the operator's intervention).
4. Transfer (coupling of two or more dies mounted in a given press or in a number of presses in such way that the parts-in-process are transferred from one die to the next immediately; there are no interruptions or delays between successive operations).

Any operation where either the feeding or the removal is made by means of some positively acting mechanism that substitutes for human intervention is called a mechanized operation.

This chapter covers the different feeding methods, and Chapter 11 covers the different ejecting methods.

Analysis of the Different Feeding Methods

In *manual* feeding the material to be processed is put into the die by hand, directly, or with some device or hand tool, indirectly. In the case of *semiautomatic* feeding, the material is placed under the punch by some mechanical means that requires the direct intervention of the operator at each press stroke. *Fully automatic* feeding does not require the attention of an operator at each press stroke; his intervention consists of periodically filling some feeding attachment with the stock or blanks to be processed. The material is then placed under the ram automatically by a mechanically actuated appliance.

Mechanized feeding methods do influence productivity not only directly (by a substantial increase of production rate in comparison with manual feeding) but also indirectly. By their use, work spoilage is reduced considerably (because of increased accuracy), die maintenance is reduced appreciably (tool life increased because of greater regularity of performance), and accidents are reduced notably (see Chapter 13).

Mechanized feeding can be provided for practically any type and size of press. However, it is always good practice to consult the corresponding manufacturer before deciding about the matter.

The correct and most adequate feeding method and means for a given press operation depend chiefly on the type of operation, the type of stock, the size and shape of the stampings, characteristics (type, design, etc.) of the dies and the available equipment (press and accessories and/or attachments), the production rate sought, and the total quantities of workpieces to be produced.

It is the type of operation that determines the fundamental feeding method. Since the stock is presented in quite different shapes in first operations than it is in second operations, it follows that the feeding methods and means are basically different for these two categories.

First Operation Tools

Chart 10-1 shows the usual feeding methods for first operation tooling.

In the case of *manual* feeding, the strip or coil is advanced between two blanking operations, by hand, locating it always with the aid of a stop provided for this purpose.

When strips are too long and/or too heavy, it is troublesome to move them, because of the excessive overhanging weight. To avoid this problem and to accelerate operations it is recommended that some simple wooden blocks, or better, some proper auxiliary tables (Fig. 10-1) be provided for these cases. Of course, the height of the table top must be adjustable in order to hold it exactly at the same height as the die-plate top, in order to keep the strip perfectly aligned. In order to adapt it to inclined presses, the table top should also be inclinable. In this way the table top and the die-plate top surfaces are maintained in the same plane. (Although such accesso-

Chart 10-1: Feeding Methods for First Operation Dies

Manual	Advance governed by 1. manual stop 2. semiautomatic stop
Semiautomatic	Advance governed by semiautomatic stop
Fully automatic	Advance performed by 1. roll feed 2. slide feed (e.g., air feed) 3. hitch feed 4. grip feed 5. transfer feed

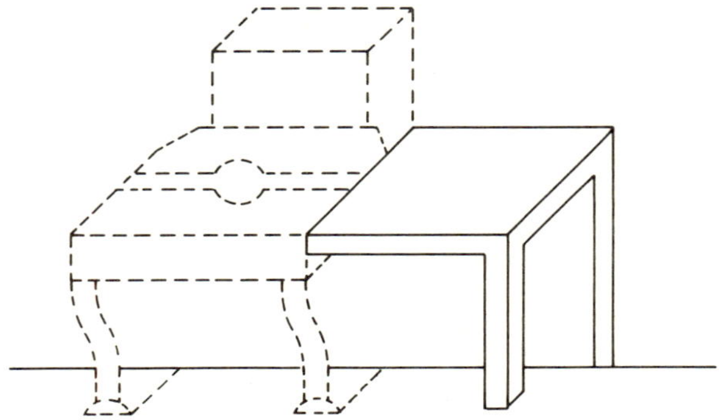

Fig. 10-1. Auxiliary table for feeding long and heavy strips.

ries may also be convenient for the strip skeleton, they are generally not required since the weight of the strip skeleton is usually not sufficient to cause a problem.)

In addition, proper guide channels should be provided in order to ease the feeding of the strip into the stock guides of the die. Such additional channels should have a funnellike configuration. They may be provided either on the strip-holding auxiliary table or on the tool itself. In the former case, of course, they must be adjustable, in order to accommodate different strip widths.

Often, instead of simple tables, the same trucks that are used to transport the strips are used to feed them into the press (see Chapter 2). These trucks then must have adjustable tops.

A very useful, yet inexpensive aid to help lift sheets and/or strips is a vacuum cup provided with a ring (Fig. 10-2). The operator wears the cup on his thumb, so that the hand is free to perform necessary working motions.

There is a limit to the strokes per minute at which a press can be fed manually for a long time and at a steady pace. According to the type of operation, size of the work and of the stock, size of the press, the operator's skill and experience, etc., the maximum attainable practical rate is about 45–50 strokes per minute.

With the *semiautomatic* method the operator does nothing other than push or pull the strip in the right direction against a special, so-called "automatic stop," which is actuated by the press. Obviously, this kind of feed is faster, about 100% more than manual feeding. Usually, with this method, it is possible to work with a continually running press, i.e., to catch almost every flywheel turn, while the manual stop, single-stroke technique allows the operator to catch only every other flywheel turn (if the press is rather fast). In other words, semiautomatic feeding has almost the same production rate as fully automatic feeding.

Fully automatic feeding is, of course, the most efficient. It offers the highest production efficiency with the fringe benefit of the best safety characteristics. On the other hand, it is not recommended for complicated, sophisticated press tools, which cause more problems than plain, simple tools.

Fully automatic feeding does not require any constant intervention or attention of the operator during the regular progress of the work; it is only required at the start. One operator can tend a battery of presses (up to three or four) since his activity consists only of providing the stock (at long intervals), of periodic inspection of the presses, and of inspecting the work. In such cases the press operator ceases to be a

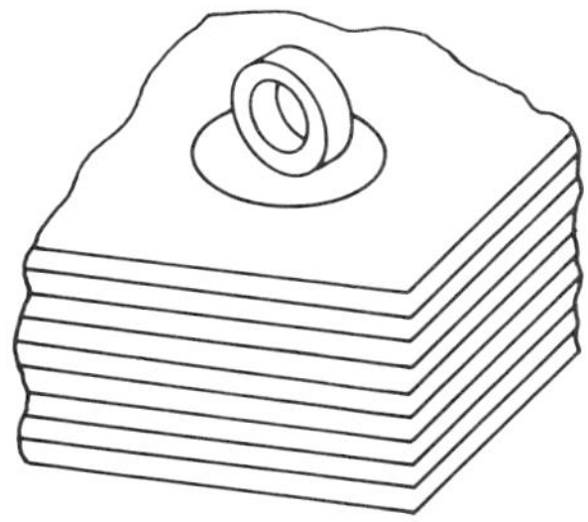

Fig. 10-2. Vacuum cup for handling sheets.

press attendant. He becomes an attendant of the press accessories or attachments, mainly of the feeding devices.

For automatic feeding, the stock is almost always used in the form of coils, but sometimes it is used in long strips. The latter give practically the same production rates as coils; however, they keep the operator busier, so that he cannot tend more than one or two presses.

Automatic Feeding Devices

There are several types and makes of automatic feeding devices for first operation dies, as will be specified below, so that practically every kind of need may be satisfied. The details that limit their adoption are few and may be summarized as follows: short runs in general, compartively large blanks in short strips, and sometimes intricate (formed) stamping shapes.

The automatic feeding devices used are practically always commercially available products; usually already attached to the presses (bolsters) when purchased. They may be applied later to existing presses by mounting on the bolster plate or even on the die-holder.

Note. A detailed description and specification of the automatic feeding devices is outside of the scope of this book. They can only be listed here. For detailed engineering information, the corresponding manufacturers should be contacted.

In the case of hand and foot presses manual feeding is usually employed, sometimes, semiautomatic feeding, but practically never fully automatic feeding. With power presses any of the three feeding methods may be used.

From both the safety and the production efficiency standpoints, of course, the fully automatic feeding methods are the best, next is semiautomatic feeding, and last is manual feeding. On the other hand, first and maintenance costs are in inverse proportion to their usefulness: fully automatic feeding devices are the most expensive, while manual feeding requires the least expense for auxiliary devices (sometimes they do not require any at all). In addition, occasionally the form, shape, or size of the components to be produced make it impossible, or at least impracticable, to use mechanically operated semiautomatic or fully automatic devices, and, therefore, for such cases manual feeding is compulsory.

In the case of automatic feeds, the press strokes per minute (spm) must bear a certain direct (inverse) relation to pitch, *L* (center distance or advance length). It is

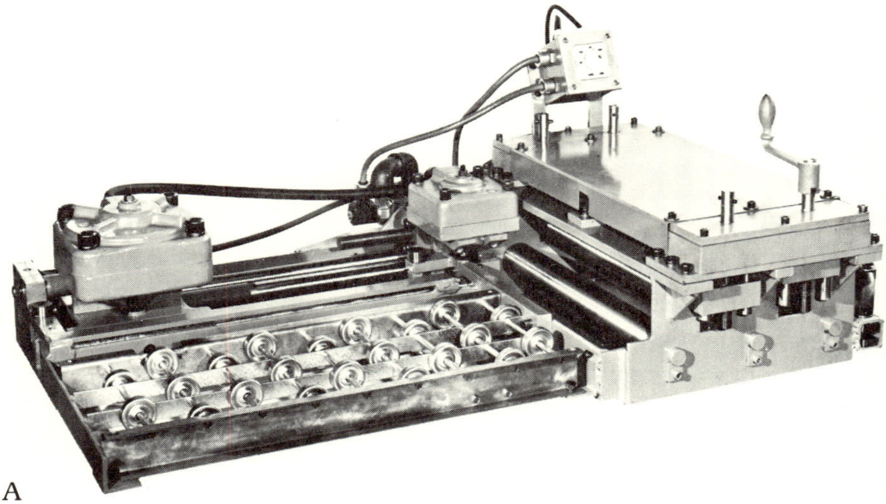

A

Fig. 10-3. Automatic air feeds. (*A, Courtesy of Jaybird Automation Inc.; B, Courtesy of American-Monarch Machine Company, Inc.*)

considered good practice not to use stock travel speeds (linear velocity, $v = \text{spm} \times L$) of more than 1000–1200 inches per minute for precision roll feed and 400–500 inches per minute for commercial feeding devices. The lower the feeding speed, the greater the accuracy of the advancement.

1. *Roll feeds.* Roll feeds used to be the most universally employed automatic feeding devices. A roll feeding attachment consists of two cylinders held together elastically by springs or counterweights, actuated by the press (ram, flywheel, etc.) through a friction or ratchet clutch, or by the tool (i.e., punch holder) during the ascent of the ram. The actuation is effected usually by mechanical means (a straight roller grip drive is preferred for short center distances and a rack and pinion roller grip drive for longer pitches), however, hydraulic or pneumatic feeding devices are also employed.

Feed length is always adjustable with micrometer accuracy, in many cases while the press is running. Also, height is adjustable for different tools. Timed power brakes are usually provided to avoid overtravel due to inertia and backward movement of rolls during nonfeeding periods. In many cases the top roll is released automatically at the start of the ram descent, in order to allow pilot adjustment. Guide rails should be provided for lateral alignment of strips. Use single rolls when the stock is in the form of coils, the stock is comparatively heavy (strong); and there is no danger of buckling. Push, when strip skeleton is weak and/or in the case of scrapless design; pull, when the skeleton is strong. Double roll feeds are used for thinner material, in order to prevent buckling, by keeping a steady tension on the strip or coil, and for strips, in order to fully utilize the stock in the case of progressive-type dies.

Roll feeds have some limitations. They are not recommended for heavy or even moderately thick stock, because feeding is then not accurate and, in addition, damage to feed mechanisms happens very easily (gears may come out of mesh with disastrous consequences for the feeding device, the tool, and even the press).

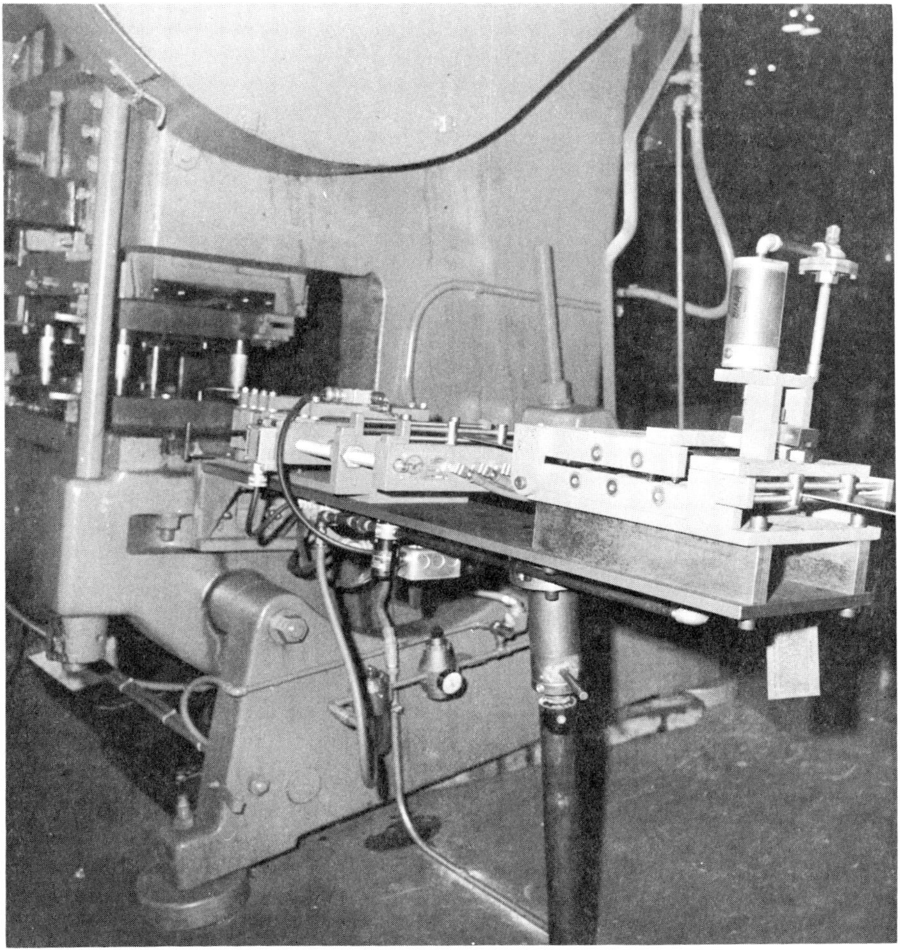

Fig. 10-3 (*continued*).

Even with thin stock, roll feeds are only moderately accurate. Tolerances in the center distances, of ± 0.005 inch or more must be allowed for them, therefore, they are preferred usually for simple blanking tools. For progressive-type tools, the use of proper aligning pilots and/or French stops is recommended.

2. *Slide feeds.* The tendency today is toward mechanical slide feeds (see Figs. 4-3 and 10-3). These feeds are actuated by the crankshaft extension of the press or they may be self-contained, independently air or hydraulically operated cam-actuated units. They incorporate a feed block reciprocating between two adjustable positive stops and a hardened pivoting blade, which engages and moves the stock. They have rather good feeding accuracy: ± 0.0015 inch. Air slide feeds are very popular today because of their excellent performance, low cost, and other advantages.

3. *Hitch feed.* Hitch feeds consists essentially of a reciprocating head that carries

Fig. 10-4. Transfer press. (*Courtesy of Waterbury Farrel, Division of Textron, Inc.*)

a gripper unit (consisting of a gripper bracket or anvil and a pivoting gripper plate or finger), which grips the stock firmly on the forward movement and releases it at the backward stroke, and a similar stationary unit. The first unit does the actual feeding, the second prevents any backward motion of the stock. Actuation of the reciprocating head is effected by means of springs compressed during the downstroke or cams attached to the punch holder.

There is greater accuracy with hitch feeds than with roll feeds because the stock is actually gripped during advance and there is no overtravel due to inertia. Tolerances can be held within ± 0.002 inch.

Hitch feeds allow higher speeds than roll feeds. They are used for rather thin and light stock (0.001–0.031 inch thicknesses). Grip and slide feeds are used for heavier stock (up to 3/16 inch or more).

4. *Grip feeds.* Grip feed is analogous to hitch feed with the difference that the gripping unit consists of pivoting shoes, instead of gripping plates and brackets. The cylinder-type grip feed (also called "die feed") operates with two pairs of gripping rollers.

5. *Transfer feeds.* Such devices are used when progressive-type dies of standard or special design cannot be used because of excessive scrap or too great depth of draw, or special shapes, or special operations are needed. In transfer feeds the workpieces to be processed are carried automatically from one work station to the next, until the stamping is completed.

Transfer feeds may be employed either in multistage dies with standard one-plunger presses, for a few stations, or in multiplunger (multiram) special transfer presses, for several stations.

The transportation of the workpieces is effected by mechanical or hydraulic "fin-

Fig. 10-5. Transfer press tooling. (*Courtesy of Waterbury Farrel, Division of Textron, Inc.*)

gers," with either a swinging or a reciprocating motion. (Transfer presses have only reciprocating gripping devices.)

Figure 10-4 shows a transfer press manufactured by Waterbury Farrel, and Fig. 10-5 shows in close-up the tooling for producing a deep drawn part, with transfer feed.

The chief advantages of these machines may be summarized in the following basic points.

1. There is a continuous positive part control by means of mechanical gripping devices perfectly timed with ram movement of the press.

2. Tooling may be provided for performing a great variety of cutting and forming operations and/or combinations of same, which are impossible to realize with standard or special progressive-type dies.

3. Very simple, and therefore inexpensive, tools are required. Job set-ups and job change-overs are effected quickly.

4. As a fringe benefit, in the case of deep drawn parts, the time factor is taken advantage of. Since the rapid subsequent redrawings are performed within seconds, the parts do not cool and the stock does not crystallize. Consequently, all intermediate annealings are eliminated.

Often, in transfer presses the first—blanking—operation is fed by means of a standard automatic feeding device properly synchronized with the transfer feed.

Feeding Accessories

In order to increase the efficiency of the automatic feeding devices, to facilitate the press operations, to improve the product quality, and to increase overall productivity, it is good practice to also use some additional auxiliary equipment, closely related to press attachments, such as stock reels, coil cradles, stock straighteners, scrap cutters, etc.

Stock reels (also called reel stands, decoilers) are adjustable pedestal stands for holding coils and unwinding them in order to feed them easily into the dies. They can have a vertical or horizontal arrangement and can be plain or automatic (i.e., motor driven); see Figs. 4-3 and 10-4. They are provided with suitable brakes to avoid overthrow or overtravel.

Motor-driven stock reels may provide a suitable constant loop between the feeding device and the reel so that the feeding device will only have to move a slack loop of material and not have to turn the stock reel also.

Coil cradles (see Fig. 4-2) serve the same purpose as stock reels, but they support the coil by means of two vertical, adjustable parallel plates. These are preferred for heavier stock and/or wider coils, while reels are better suited for thinner gages

Fig. 10-6. Coil lifter and positioner. (*Courtesy of Bushman Equipment, Inc.*)

Fig. 10-7. Scrap rewinder. (*Courtesy of American Feed Co.*)

and/or narrower ribbons. (In case of doubt, use reels; they usually give less trouble.)

Stock straighteners are devices that eliminate waviness, curvatures, undulations, kinks, "sets," etc., in strips or coils, thus preparing them in perfectly straight condition for the same operations. Also, these devices may be power driven, for comparatively severe conditions (see Figs. 4-3 and 10-4). Usually they are combined with the feeding device, sometimes also with the coil cradle or stock reel and/or the oiler and wiper.

Stock straighteners are also often used in cases where straightness is not an essential condition for the product. In almost every case it is the press operation that demands straight stock, in order to avoid several types of mechanical problems, e.g., safety in feeding and accuracy of pitch length.

A *coil lifter and positioner* (Fig. 10-6). A coil lifter and positioner is a useful and practical device for handling and storing efficiently, large and heavy coils or rolls.

Roller levelers. Roller levelers are sometimes used to take out the set from curved coils and to relieve internal stresses, in order to make the stock both straight and more ductile. Usually, they are located between decoiler and die (see Fig. 4-5).

Scrap winders (rewinders). In the case of comparatively thin stock and greater accuracy requirements, it is good practice to provide motor-driven stock reels for winding up the strip skeleton, for easier, more efficient scrap handling (Fig. 10-7). Obviously, such accessories can be used only if the strip skeleton has sufficient strength, otherwise scrap cutters are recommended.

Scrap cutters. In many cases, it is preferable to avoid the trouble of handling the strip skeleton; therefore, the press—or the die—is equipped with a scrap cutter (chopper). This consists of a simple shear located at the exit side of the tool and adjusted so that at each ram stroke the scrap (coil skeleton) is cut into short lengths exactly corresponding to the center distance. Such short pieces of scrap skeleton are very easy to handle and transport; in addition, scrap cut in this way has a higher commercial value (see Fig. 4-3).

If the feed length (i.e., center distance) is too short, the scrap will consist of very small pieces that are difficult to handle. For such cases, the scrap cutter may have intermittent motion, skipping one or more strokes. In this way, the scrap is cut to lengths that are double or triple the feed length.

Tables for sheet handling. Figures 10-8 and 10-9 present practical, actual applications of special heavy-duty tables for handling metal sheets, large stampings, heavy strips, etc. These tables are produced by the American Manufacturing Company. The top of these tables is easily adjustable (by means of foot pedal control) in such way that the top level of the sheets or parts is always at a convenient and easily reachable position for the press operators. (See more information about sheet handling tables in Chapter 2).

Oilers and wipers. Oilers and wipers are discussed later in this chapter.

Sheet separators. In the case of stock worked in the form of rather wide strips or in sheets, magnetic sheet separators (fanners) may be employed successfully, making stock catching easier and thus faster. Essentially, such devices consist of permanent magnets that temporarily magnetize the sheets. These are subsequently separated automatically since the same magnetic poles repel each other.

Signals. When an operator is tending simultaneously a battery of presses, it is strongly recommended to provide the press with some acoustic device which will

Fig. 10-8. Table for sheet handling. (*Courtesy of American Manufacturing Company, Inc.*)

Fig. 10-9. Table for sheet handling. (*Courtesy of American Manufacturing Company, Inc.*)

advise the operator as soon as the stock is finished in a given press and/or when any kind of irregularity or problem arises in the working of a press. In these cases, of course, the press must be stopped automatically at this time in order to avoid needless motion and possible problems.

Feeding of Second Operation Dies

As mentioned before, from the productivity standpoint multioperational dies, especially the very popular progressive-type dies—where at each press stroke a complete, finished stamping is produced—are to be preferred to single-operation dies—where at each press stroke only one operation of the production cycle is performed so that several separate operations are needed to complete a stamping. However, there are quite often situations which require or at least make it advisable or convenient to use the single operation production method—one first operation tool to produce the blank and as many second operation tools, as needed. A few such situations are listed here.

1. When the same primary blank or the same partly processed stamping is used for various similar products of slightly different characteristics.

2. When a standard progressive die would have too many stations, consequently, the tool would become too large and long and there is no available press in the shop with a sufficiently large table.

3. When the duration of the different phases of production vary too widely (quick cutting actions mixed with slow forming operations), so that the overall duration of the whole production cycle would be unnecessarily long.

4. When in the production cycle some assembly operation is included (a separate component must be incorporated into the main stamping before the last operations).

5. To keep the first cost and maintenance costs of tooling low. (A progressive die costs, as an average, four to five times as much as the entire set of corresponding single operation dies.)

6. When the planned quantity of stampings to be produced is limited, so that there is no ecnomic justification for spending too much time and money for the more-sophisticated and expensive multioperational tooling.

Basic Concepts

There is no such thing as a universal feeding device or feeding method. Every part, every setup must be studied individually, and the most appropriate feeding method selected. The selection is based on the following characteristics:

1. Part: size, shape, stock thickness, thickness/size ratio, material, dimensional tolerances
2. Operation: cutting, forming (bending), drawing
3. Feeding method: direct, indirect, semiautomatic, fully automatic
4. Production rate: independent or dependent on operator's working rhythm
5. Application: individual or adjustable (universal or multipurpose)
6. Production volume: small, medium, high
7. Ejection of finished stamping and scrap: independent of feeding device; incorporated into feeding device
8. Press body construction: vertical, inclined, not a determining factor

Chart 10-2 lists the different feeding methods used in second operation cutting and forming tools.

Manual feeding. Correct safety practices forbid direct manual feeding because of the high probability of an accident occurring. (See Chapter 13.) Only in extreme, special cases may direct manual feeding be used, such as in second operation blanking from scrap material or from unit stock of nonuniform shape, trimming of castings or otherwise prepared nonuniform blanks, or finish forming or progressive forming of complicated parts. For such cases, some auxiliary tables around the dies (Fig. 10-10) similar to those for first operation manual feeding of strips, make possible a production rate increase of about 20–25%.

In the case of direct feeding, the operator's hand or fingers should never reach into the danger zones of the dies. This is achieved by means of auxiliary safety hand tools widely used in press working of metals. They are simple pliers or tongs, and vacuum, pressure, or magnetic lifters. Usually such hand tools are made from special light metal alloys so built that when the press operator by mistake or distraction trips the press while the tools are still in the interior of the die, there is no damage to the die nor injury to the press-operator. The tool points simply disintegrate.

Chart 10-2: Feeding Methods for Second Operation Dies

Manual	1. Direct, without feeding aids 2. Direct, with grippers, tongs, vacuum cups 3. Indirect gravity chutes follow chutes plunger feeds push or finger feeds reciprocating slides rotating feeds movable dies pivoting swinging revolving reciprocating *Note:* possibilities of magazine and hopper feeds exist for all methods, except for the chutes and plunger feeds
Semiautomatic	Power-driven swinging arms reciprocating slides movable dies Dial feeds
Fully automatic	Mechanical feeds Power-driven reciprocating slides with magazine movable dies with magazine dial feeds with magazine
Multioperations	Interoperational handling

In Figs. 4-4 and 13-18 such safety tongs are clearly visible at the front of the presses. Figure 10-11 illustrates the working principle of such a feeding method.

In some exceptional cases it is possible to manually feed the dies directly without the need of any additional device or aid. These are the cases where the parts are quite large and only a comparatively small area of the blank is involved in the press operation. Consequently, the blank may be held safely outside of the die, away from the danger zones, during the operation.

For the best safety and productivity conditions, *indirect manual feeding* methods are to be preferred. With very simple, inexpensive devices, e.g., a simple chute or slide to move the parts to and from the die, which often can be homemade in a couple of hours by semiskilled workers, such as a die setter, production rate increases of 50% or more are easily effected.

The best procedure consists, of course, of designing and building the feeding device together with the die. However, in many cases it is quite possible to add the feeding device afterward, when the die is already in production.

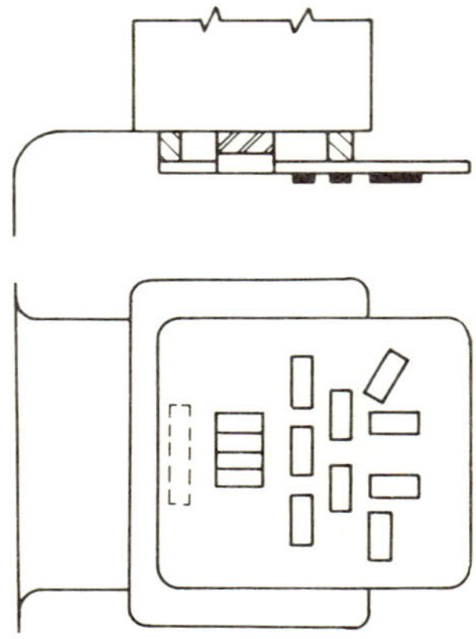

Fig. 10-10. Auxiliary table for facilitating manual feeding of secondary operation dies.

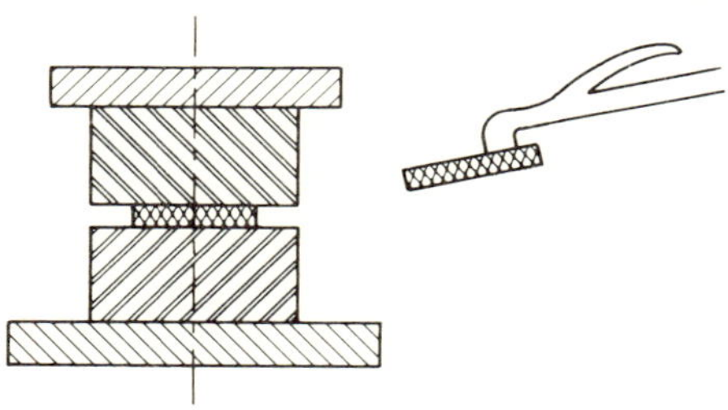

Fig. 10-11. Safety tongs used for direct feeding.

The most usual *indirect manual feeding* methods are gravity chutes (troughs), follow chutes, plunger or push feeds, reciprocating slides, movable dies (sliding, swinging, pivoting, revolving).

Gravity feeds. The simplest, least expensive, indirect manual feeding method takes advantage of gravity. This kind of feed consists of a simple inclined chute with side guides between which the blanks to be processed are conveyed to the die. The blanks slide down automatically, one at a time, into the die when the operator feeds them from above, well away from the die and its danger zones. The blanks are stopped there by pins or rails or other suitable means.

It is good practice to provide a suction breaker—wire rails on the chute bottom—in order to avoid sticking of oily blanks (Fig. 10-12). There are several possible practical designs for such suction breakers. Figure 10-13 presents the simplest, self-contained design. There are a few (two or three, according to width) rails formed in the sheet metal in the chute bottom. In Fig. 10-14 the rails are extra items fastened to the chute bottom. The design in Fig. 10-15 is employed in case of heavy stock. Here, strategically located steel balls, held in such way that they are free to rotate, facilitate the sliding of the workpiece blanks.

Gravity chutes are intrinsically more efficient than direct feeding methods, because it is much quicker to put a blank into a chute than directly into the die. The inclination of the chute depends chiefly on the weight of the workpieces; heavier stampings need less inclination and vice versa. By partially covering the chute, an

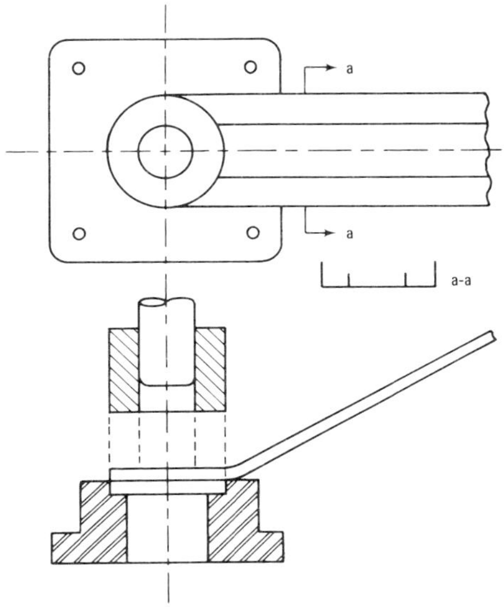

Fig. 10-12. Feeding chute for gravity feeding.

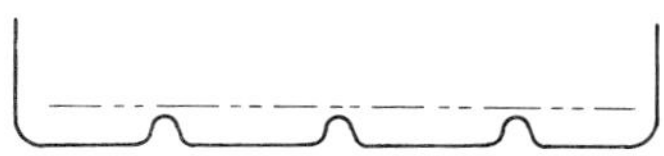

Fig. 10-13. Suction breaker rails for inclined chutes.

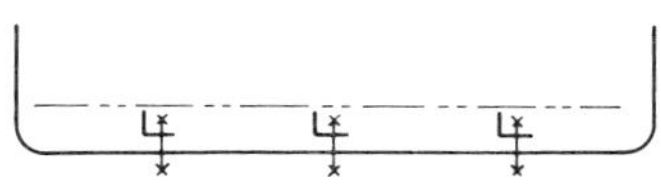

Fig. 10-14. Improved suction breaker rails.

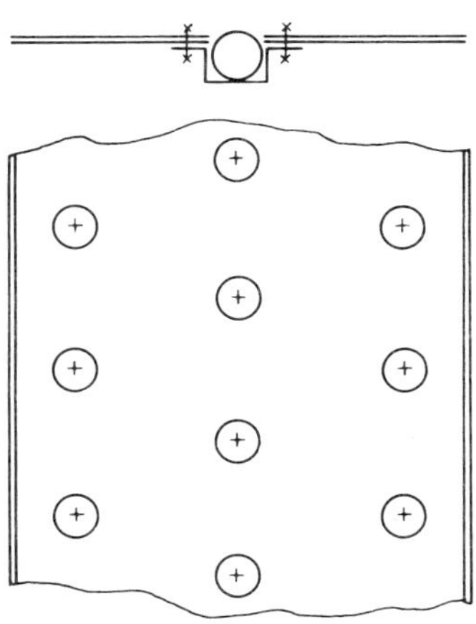

Fig. 10-15. Chute design for heavy stock feeding.

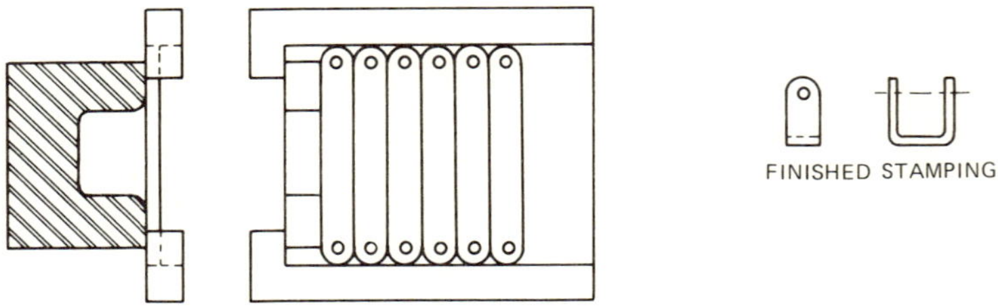

Fig. 10-16. Follow feeding arrangement.

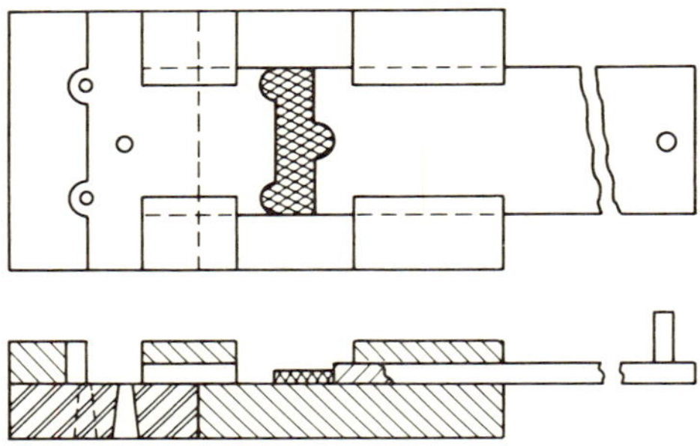

Fig. 10-17. Push feeding method.

additional advantage is gained: two blanks cannot be fed simultaneously into the die.

Gravity chutes may be used for either upright or inclined presses. In the latter case, the whole setup becomes simpler and more natural, and, in addition, the ejection of the finished workpieces becomes much simpler. This combination (gravity chute plus inclined press) gives the simplest, least expensive, and fastest method of indirect feeding.

Follow feeds are used when the shape of the blanks does not allow gravity feeding or when there is a need to control the blank even after it has been put into the feeding device. (In the case of gravity feeding, it is not possible to do this.) The blanks are placed in correct orientation into the horizontal chute and the operator feeds them into the die by keeping the chute full and pushing the extreme outside blank. Thus, one part pushes the parts ahead of it into position (Fig. 10-16).

Push feeds may be used when the shape of the blanks does not allow either gravity or follow feed. Push feeds also serve for the proper, correct, pre-determined orientation of the blanks. Such feeds are essentially guided plates having at the inner end an open half-nest for the blank. The other half of the nest is stationary (in the interior of the die) and also constitutes the stop (Fig. 10-17). The part to be processed is put into the push feed outside, at a safe distance from the danger zones of the die, and then the plate is used by the operator to push the blank into the die.

Reciprocating slide feeds (shuttle) are a modification of push feeds. The difference between them is that reciprocating slides have full nests, instead of partial (open) nests. Therefore, they can be employed not only for feeding, but also for removing the processed stampings after the operation (Fig. 10-18).

There are some special cases, with quite different characteristics, where the die plate is not stationary, but may be withdrawn from the tool. One of current methods

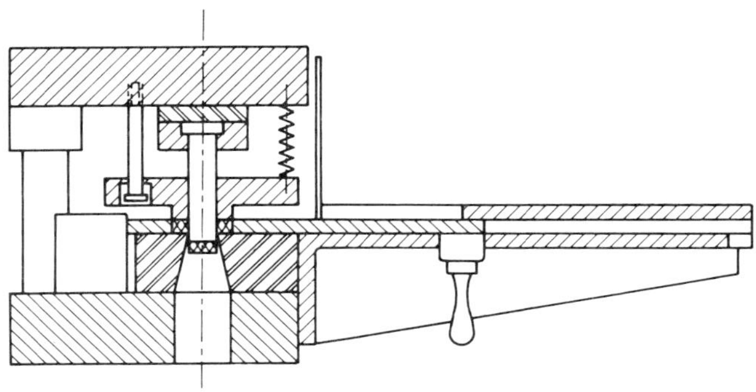

Fig. 10-18. Reciprocating slide feed.

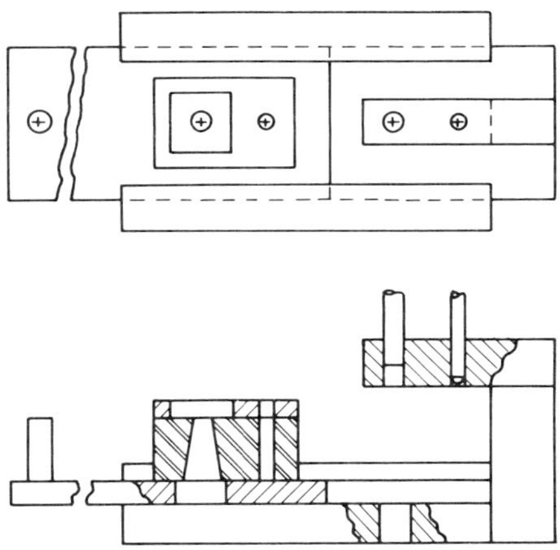

Fig. 10-19. Movable die on reciprocating slide.

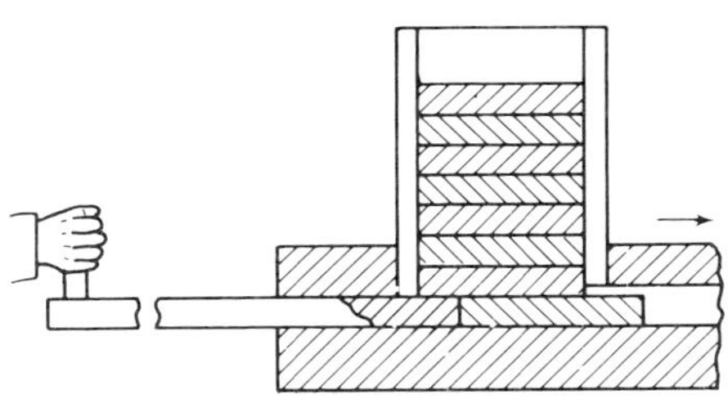

Fig. 10-20. Slide feed with magazine.

for *movable dies* (or more correctly, movable die plates) consists of mounting them on reciprocating slides or guided bolsters (Fig. 10-19). Other arrangements for movable dies utilize pivoting arms, revolving tables, or swinging tables.

Sometimes the efficiency of the indirect manual feeding method (push, reciprocating slide, etc.) is improved by adding to the equipment suitable magazines which are periodically refilled, by the operator, with blanks to be processed (Fig. 10-20). There are several typical basic designs for such magazines. Figure 10-21 presents a split tubing shape. In Fig. 10-22 is shown a magazine formed by four strategically placed round rods. Figure 10-23 is a skeleton magazine for square parts. Other designs utilize structural steel bars.

If for some special reason it is impossible or economically inconvenient to employ any of the above-specified indirect manual feeding devices, then direct manual feeding must be applied always using, however, some gripping tool, as mentioned above.

Semiautomatic feeds mean a considerable increase of production rate, sometimes up to 100% or more in comparison to manual feeding. Semiautomatic feeding devices are essentially evolved from the manual, indirect feeding devices by converting them into power-driven units. The power for the actuation is mechanical in the majority of cases (directly or indirectly from the press main shaft), although quite often pneumatic or hydraulic power and sometimes electric power is used. So, we have (a) mechanical push feeds, (b) reciprocating slide feeds, (c) swinging arms, (d) reciprocating sliding dies, and (e) rotating table feeds, these latter better known as dial feeds.

All these semiautomatic feeding devices have a similar working principle, with slight variations. In all cases the operator places the primary workpieces to be stamped one at a time in a nest or recess that is well outside of the tool and then actuates the press. The rest is performed automatically by the press and the feeding mechanism.

The mechanical push feed pushes the blank into the die and holds it there during the stamping operation. During the upstroke of the ram the pusher is automatically brought out of the die in order to be ready to receive the next blank. The finished workpiece is evacuated from the die in any of the standard or special methods. The slide feed works analogously, with the only difference being that the slide can be utilized also for the unloading of the finished workpieces.

In another type of semiautomatic feed, swinging arms catch the workpieces placed in a given location and deliver them to the interior of the tool and subsequently remove them.

In the case of movable dies and dial feeds, the workpieces to be processed are placed in the corresponding nests which are mounted on the dies.

Dial feeds sometimes allow the possibility of performing a few operations simultaneously (progressive stamping) thus increasing production efficiency.

Of course, every feeding device must remain still ("dwell") during that portion of the ram downstroke that corresponds to the performing of the stamping operation. In addition, proper mechanical or electrical interlocking safety devices must be provided in order to avoid the actuation of the press if the feeding mechanism is not in its correct closed position.

In some cases, the operator does not actuate the triggering of the press, but only the feeding mechanism. When this has moved to its correct closed position, it actuates automatically the press clutch.

The feeding cycle may be intermittent as described up to this point or, as happens quite often, the press may run continuously and the operator must catch the right moment to put the workpiece into the feeding device. The feeding device then carries the workpiece into the die automatically, in perfect synchronization with the ram stroke. The feeding device movement is usually positive in both directions, with the exception of push feeds.

With power-driven semiautomatic feedings, mounted in a continously running press, the *speed* (strokes per minute) of the press should be correctly synchronized to

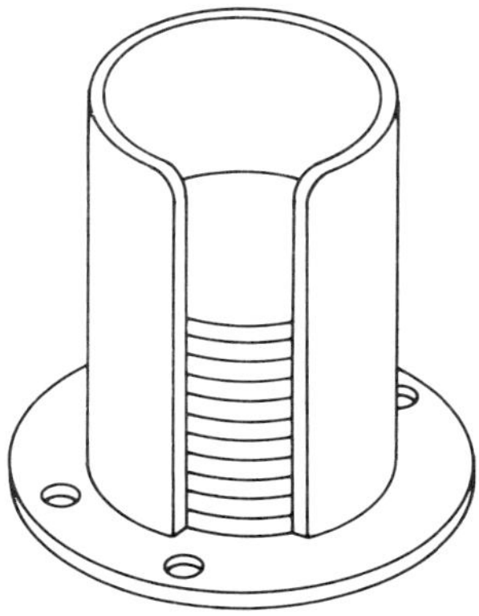

Fig. 10-21. Split tubing magazine.

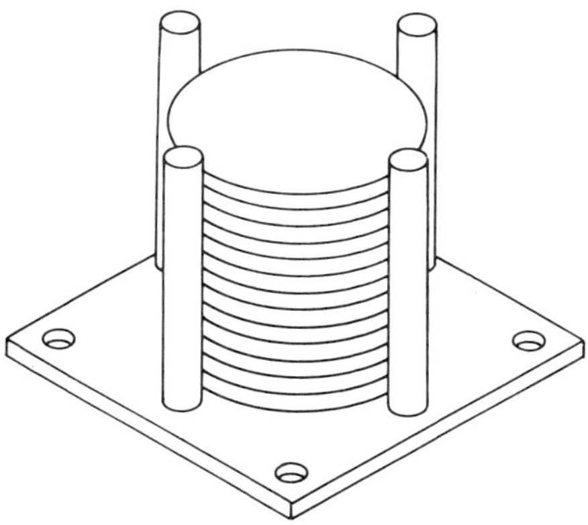

Fig. 10-22. Magazine made with steel rods.

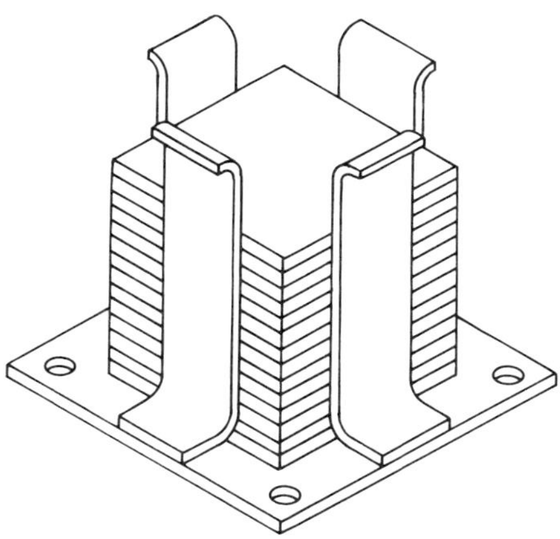

Fig. 10-23. Magazine made with flat steel bars.

the comfortable working rhythm of the operator. Too slow a press means an unnecessarily low production rate and production efficiency. On the other hand, too fast a press may be a drawback in such cases, as has been found repeatedly in actual practice. In fact, an operator can keep up the feeding of a semiautomatic mechanism only when the press runs at a reasonable speed. This may vary over fairly wide limits among individuals and even for the same individual between different hours of a working day. Let us take 45 strokes per minute as a good average speed. If the press speed is increased to, say, 50 strokes per minute, the operator then may be physically and mentally unable to catch each press stroke, but only about every other stroke. So, a faster press actually gives a much lower hourly output. Consequently, variable speed drives for presses are of a great help for adjusting the press speed to its most efficient value for each job, at each particular time. Thus the presses can be run continuously with notable advantages of production efficiency increased and reduced press wear and maintenance costs.

Indexing is a variation of dial feeding with the difference that indexing is a combination of semiautomatic and fully automatic operation. The workpiece is held on a revolving holding device and the press is tripped. At each press stroke the work-holding disk makes a rotation through an angle of a determined number of degrees. The press automatically makes the required number of strokes to complete the stamping and then is stopped automatically. The operator then removes the finished workpiece and a new production cycle is started. Sometimes indexing is done manually by changing positions with a handle. Obviously, this method has a comparatively low efficiency.

Fully automatic feeds are the next step to improve productivity by changing the working method. First, the actuation of the feeding slide or table is not intermittent and controlled by the press operator, but is always continuous. The feeding device is actuated directly by the press, through attached linkages or air cylinders, at each return stroke of the press ram. In addition, the setup is provided with some supplementary accessory (magazine, hopper, chute, drum or belt feeder, etc.) which must be kept filled by the press attendant with blanks to be processes. In this way the feeding devices are converted into fully automatic feeds and the operator is completely freed from keeping up with the rhythm of the press, and his or her performance is practically limited to material handling, i.e., filling the magazines at long, loosely predetermined intervals, disposing of finished stamping lots and scrap, and occasionally inspecting the stampings. The magazines, hoppers, chutes, etc., are replenished without the necessity of stopping the press.

For fully automatic feeding of second operation dies magazines are preferred as an auxiliary accessory, because of their simplicity and low cost. (They are described on pp. 147–148.) The blanks are stacked vertically in the magazine and before each press stroke the feeder introduces one blank into the die.

Fully automatic feeding devices for second operations may be classified as follows: automatic mechanical feeders, slide or shuttle feeds, plunger feeds, and dial feeds. Mechanical feeders ("iron hands") always take the top piece of the blank stack, while all the other types of feeding devices feed the flat blanks from the bottom of the stack in the magazine.

The automatic actuation of the slides may be performed mechanically (through linkages operated by the press or by means of cams incorporated into the dies) or controlled by the press (through cams) pneumatically, hydraulically, etc. (see Fig. 10-24).

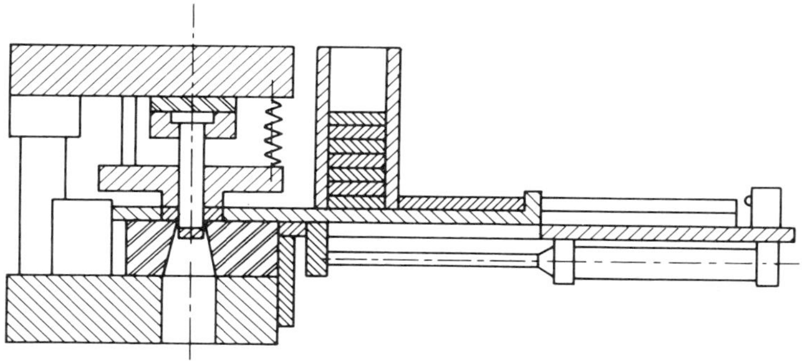

Fig. 10-24. Pneumatic fully automatic feeding.

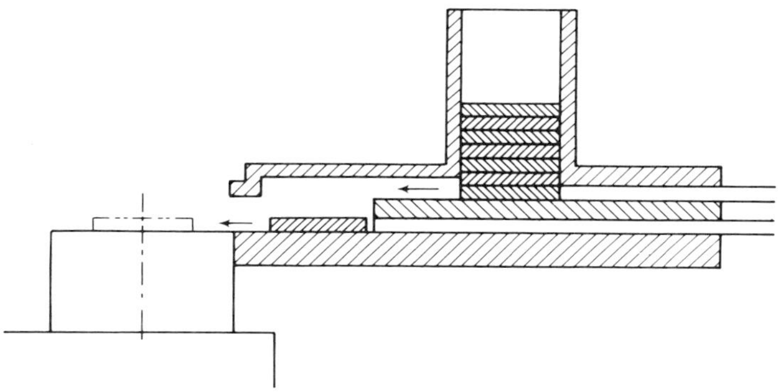

Fig. 10-25. Double pusher feeding device for long travel.

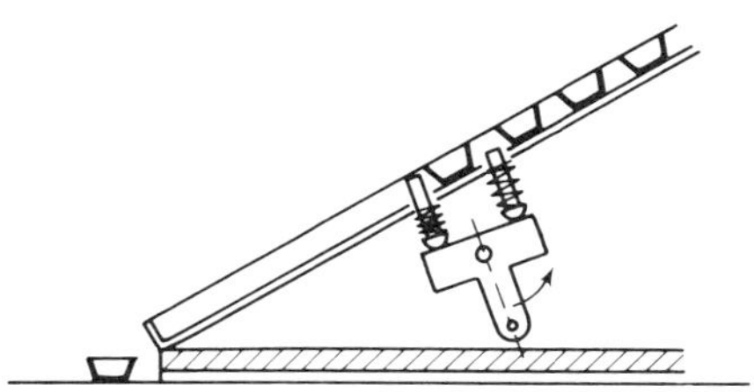

Fig. 10-26. Feeding chute with rocking escapement arrangement.

In those cases where the feed travel is too long, it may be divided into two halves (Fig. 10-25). The first plunger pushes a blank halfway; the second plunger then pushes it into the die. The two operations are performed simultaneously, so there is always a blank that is being pushed halfway while another is being pushed into the die nest.

In addition, there are also gravity chute feeds with fully automatic features. These are inclined chutes provided with a rocking escapement (intermittent shop) timed with the ram movement, and so arranged that at each press stroke only one blank is released into the die, regardless of the number of blanks in the chute (Fig. 10-26). Another type of single-piece feeder consists of a gravity chute feed integral

Fig. 10-27. OBI press equipped with vibratory bowl feeding device and dial feed. (*Courtesy of V & O Press Co., Inc.*)

with progressive (or simple blanking) dies where the blank moves at a right angle to the strip movement direction.

Figure 10-27 shows a V & O Press Company machine equipped with a completely automatic feeding arrangement. This consists of a dial feed mounted on the press table and a vibratory bowl feeding device which automatically sends the blanks to be processed through a chute to the dial feed.

Finally, it should be noted also, that an indexing operation may be converted into a fully automatic operation. This is done by adding a mechanical feeder which automatically loads the blanks into the die and automatically removes the finished workpieces.

Fully automatic feeding is entirely independent of the operator's working capacity (see item above regarding speed of semi-automatic operations), so much higher press speeds may be used; of course, these speeds must be compatible with the mechanical and other requirements of the operation and the feeding action. As mentioned above, fully automatic feeding devices have another advantage which consists of the fact that they always guarantee correct location of the blanks in the nests; consequently, there are no miscuts and the tools last longer, often up to a 100% increase in tool life is obtained.

Interoperational Handling of Stampings (Integrated Production Lines)

With very rare exceptions, a stamping is always produced by more than one operation. Even the simplest components may have been subjected to at least two separate operations: punching and blanking. There are many cases where the workpiece is finished only after six to ten operations, or even more. Thus, there arises the problem of efficiently handling the stampings between operations. (Since this activity has somewhat the character of "feeding," it is dealt with in this chapter.)

Interoperational handling of stampings customarily may be carried out in three ways: manual, semiautomatic, and fully automatic. In addition, there is a fundamental difference between the case of single press (common for all the operations) and chains of separate presses (one for each operation or for every couple of operations).

Common Press

Manual handling. It has been found that in the case of progressive press processing of formed stampings, about a 10–15% increase in production rate is easily obtained if all or at least several of the forming tools needed for a given workpiece are united in one set and worked progressively. The partially worked pieces are transferred by hand from one station to the next. Progressive drawing offers an additional fringe benefit due to the time factor, i.e., because of the brief intervals between steps, the metal keeps its ductility longer (less cold-working hardening, less crystallization) and offers better drawability conditions.

Semiautomatic handling. Semiautomatic handling of stampings with single press arrangement is not customary.

Fully automatic handling. Fully automatic handling may be effected in three different ways: (1) progressive dies, where the strip skeleton automatically carries the preprocessed stamping to the next station (see Chapter 3); (2) transfer dies, where the tool is equipped with automatic transfer devices; (3) transfer presses, where the press is equipped with automatic transfer devices.

Separate Presses

Considerable time, labor, and handling may be saved if the stampings are conveyed *directly* from one press to the next for the subsequent operation.

Manual handling. Even the simplest planned arrangement can bring significant production increases. Place the equipment (presses and workholding tables) so that the operator of the preceding operation can deposit the manually removed workpieces of his press on the table within easy reach for the next operator.

If the distances between two subsequent presses in the production line are too far away or their orientation is such that there is no convenient place within easy reach for both operators, then very simple (usually homemade), inclined (open or semiclosed) chutes or troughs may be employed for transferring the workpieces (Fig. 10-28). if the disposition of the job does not permit the use of gravity chutes, then, for the same purpose, suitable, elevated conveyors may be provided. These conveyors may be stationary or movable for greater flexibility; in either case, they should be adjustable between certain height limits. Sometimes, a common belt or roller or slat conveyor is used for a whole production line (Fig. 10-29). As can be seen, the con-

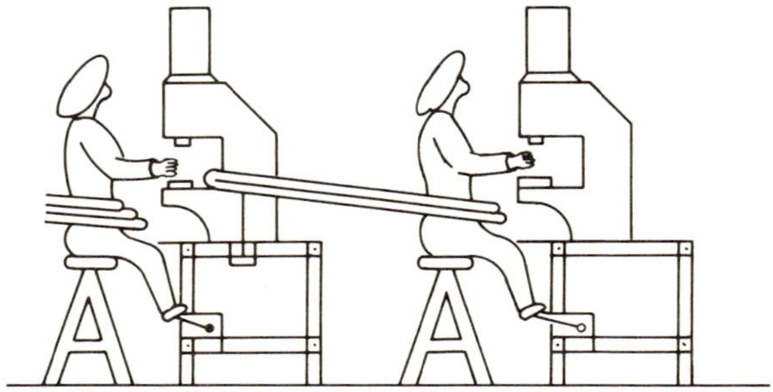

Fig. 10-28. Manual transfer of parts from press to press.

veyor belt may be provided with strategically located deflectors or sweeps that act in such way as to deliver the parts at a place within easy reach of each operator.

By eliminating the rehandling necessary when stampings are let drop into tote boxes from which the next operator must take them out one by one, considerable handling time and costs are avoided.

Semiautomatic handling. Semiautmatic handling is analogous to manual handling with the following difference: In semiautomatic handling the workpieces are removed from the first press automatically and are transported automatically with chutes or conveyors to a convenient location for the next operator to pick them up and feed them into his or her press. In other words, the operator of the first press does not take part at all in the subsequent handling process. Thus, his or her output is higher, because he or she is dedicated only to the loading and actuation of his or her own press.

The simplest arrangement in such cases is made so that the ejection at the first press is at its rear which faces the front side of the next press. The conveyor usually deposits the workpieces on a table or a flat box, possibly with such an inclination as to facilitate the catching of the workpieces.

Fully automatic handling. The next step consists of automating the production line completely. This is achieved by arranging the loading of one press in such way that the workpiece coming from the preceding press, carried by a gravity chute or conveyor, is introduced directly into the die. This must be done in perfect synchronization with the ram-stroke movement; in the case of single stroke operation, the starting of the press must be done once the workpiece is correctly located in the nest. Of course, suitable interlocking and protective devices must be provided, in order to avoid problems due to misfeeds or any other causes. If there is a problem with any of the subsequent presses, for any reason, the preceding one(s) must be stopped immediately and automatically.

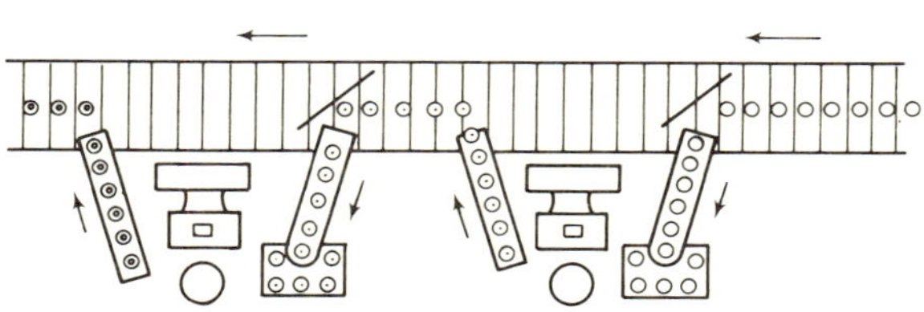

Fig. 10-29. Transfer of parts by means of a continuously running conveyor.

In some exceptional cases, e.g., in automotive production, industrial robots are employed for material handling, assembly operations, etc.

Hints for Feeding of Press Tools

There is more to efficient feeding than simply providing a good feeding device. Especially in the case of manual and semiautomatic feedings, more must be done. With first and second operation tools the convenient and proper locating of the stock is of utmost importance. It must be placed on adjustable tables and/or shelves or in bins, tote boxes, trays, etc., near the tool and within easy reach of the sitting operator. In case of second operations, the blanks or parts always should be put in the same place and that place should be chosen carefully. The operator should not have to lift or carry strips unnecessarily; he or she must be required to do the shortest and least fatiguing movements (proper "time and motion study" technique can, and should, be applied successfully). Only in such a way can the operator perform his or her task comfortably and help maintain efficient production and a high output.

A good supply of material must be kept on hand so that the operator can keep working steadily without having to leave the press at all, or only after a very long interval, and even in those cases, only for a short distance because in an easily reachable location a considerable amount of stock is held in readiness. An insufficient supply of material, or even a seemingly insufficient supply, can make the operator nervous, uneasy, slightly worried; thus, he or she is distracted and his or her production efficiency is lowered.

In the case of small components, it is good practice to keep them in comparatively large and conveniently located containers from which the operator takes a small quantity and puts it in a small box, which in its turn is utilized for the regular work.

With any type of manual feeding, direct or indirect, and even with semiautomatic feeding, a two-operator team may effect considerable labor savings and a very high press utilization factor. For example, in the case of press assembly work (riveting, crimping, curling, forming, etc.) one worker does the preparatory manual joining of the components and the second worker operates the press. In this way each operator works easier and quicker, and the overall efficiency is greatly improved.

Lastly, feeding mechanisms must be subjected to the same kind of care and preventive maintenance as presses and other pieces of equipment. The same concepts and procedures stated in Chapter 4 for presses must be adopted also for feeding devices (and, of course, any other auxiliary devices), with reasonable differences according to their particular characteristics.

Stock Lubrication

Stock lubrication is a quite controversial subject. If you wish to start an argument in a press shop, then just simply state your opinion about stock lubrication. Shortly you will find someone who will fervently advocate a different, sometimes even an opposite, opinion.

Many years ago it was still possible to find in some press shops people who were against stock lubrication, considering it superfluous. However, today practically everyone agrees that stock lubrication is not only not useless, but is definitely useful.

Some experts even think that lubrication of stock is highly essential to efficient production of metal stampings and that it has almost the same importance as tool design and tool building in forming and especially in drawing operations. Both experimental research and extensive actual practice have demonstrated several advantages that are achieved by the use of the right kind of stock lubricant—applied correctly, in the right quantity, and at the right time—during press operations. In any kind of die work there is:

improvement of product quality, especially surface finish

easier operation with less stamping effort

cooler machine operation

considerable increase in tool life (prevention of wear due to abrasive action): (1) more blanking and piercing between resharpenings; (2) in forming tools the active surfaces maintain their finish longer (scoring due to "pick up," "galling," "cold welding," "firing up," and "loading" is avoided)

reduced tool maintenance

decreased danger of tool breakage

reduction of rejects in drawing and forming, eliminating wrinkles, scratches, and other abrasion marks, as well as tearing of material and uneven shells

increased production rate

deeper draws possible at each operation and, consequently, elimination of some intermediary redrawing operation

cleaner operations which facilitate handling in second operations and inspection (quality control).

In conclusion, proper stock lubrication means increased output and, consequently, improved production efficiency.

In drawing operations the lubricant creates a film between the stock and the tool members thus allowing the stock to slip easily between the active members of the tool. In this way, the friction between the working surfaces of the tool and the stock which is being processed is reduced and, therefore, the heat produced by the operation is greatly decreased and the tendency toward metal pick up and scoring is considerably reduced or even eliminated.

The type of lubricant has a great influence on results. Always consult some reliable supplier *and* make exhaustive tests in order to find out for every type of operation and stock which lubricant gives the best results. For every type of operation and stock there is a lubricant that is best, but there is not one type of lubricant which is equally good for all press operations and all types of stock. There is no such thing as a "universal" lubricant.

Chief Requisites

The chief requisites which must be satisfied by a good lubricant are listed in Chart 10-3.

Chart 10-3: Chief Requisites for Stock Lubricants

Satisfactory performance, i.e., must have good film strength and high viscosity at elevated temperatures.

Good adherence to the working surfaces in order to control the friction between work and tool (friction should not be too high nor too low).

High resistance to heat and to pressure (stamping forces).

Low surface tension for quick spreading.

Can be applied easily.

Washability, i.e. easily removable.

Absence of corrosive staining and rusting of metallic surfaces, i.e. stock, tools, machines, etc.

Solubility for easy mixing.

Absence of toxic effects for people (toxic fumes, skin irritants, dermatitis causing ingredients, etc.).

Absence of disagreeable odors and feel.

Should not produce alkaline deposits.

Stability in storage, i.e.. it should not become rancid or freeze, etc.

Weldability (occasionally), i.e., unwashed workpieces can be resistance welded, satisfactorily

Uniform quality, i.e., once the correct brand and mixture is found, the supplier must always deliver it with the same characteristics as the sample with which the tests have been made.

Selection Proper

Since selection of the most suitable lubricant is still an art, and not yet a science as it should be, there is a wide difference of opinions in this respect. At any rate, there seems to be a general consensus on a few points to take into account in the selection of the correct lubricant. The selection is based on the following factors:

1. type of metal to be stamped, and its physical properties, thickness, age, etc.
2. type of operation
3. severity of operation
4. speed of operation
5. tool design
6. surface finish conditions of the stock and of the active tool members
7. shape of workpiece.

It must be emphatically stated that only through practical trials is it possible to determine the right lubricant in each case, changing, if necessary, the components or ingredients of the lubricant compound and/or their proportions. Since lubricant costs are comparatively low, always select the best available lubricant for any given job. Also, for obvious reasons, try to use as few types of lubricant as are compatible with correct and satisfctory operations.

Never leave the selection of the lubricant to the foreman, supervisor, or operator. The widespread practice of letting the operator use the first lubricant he has at hand should be strongly discouraged. The lubricant to be used for each operation should be specified by the Engineering Department and posted on the process sheet.

Always keep in mind that, especially in case of severe forming operations, the lubricant quality may be the deciding factor between the success and failure of the operation.

Cutting Operations

It seems that stock lubrication does not create problems in blanking, punching, and other kinds of cutting operations. At least this is implied by the lubricant manufacturers who concentrate their promotional efforts chiefly on suitable drawing lubricants. This should not mislead one into believing that for cutting operations *any* lubricant is satisfactory. Even if it must be recognized that forming operations, owing to the greater plastic flow of metals involved, are more critical and require closer attention, this does not mean that cutting operations cannot cause problems, too, if lubrication is not properly considered.

The selection of the correct lubricant for a cutting operation is governed principally by the type of stock. (See Chart 10-4 for lubricants recommended for average conditions.)

Chart 10-4: Lubricants for Cutting Operations with Press Tools

Steel: mineral oil, sperm oil, lard oil, soluble oil, or sulfurized oil. A heavier lubricant is suggested for steel of low shearing strength than for stock of high shearing strength.

Copper, brass: the same as for steel, sometimes with an addition of about 10% kerosene or soap diluted in water.

Aluminum: kerosene, sometimes with the addition of 10% mineral oil or soluble oil (emulsion).

Zinc, lead, tin: soap applied hot on the stock or heavy mineral oil.

Magnesium: mineral oil.

Plastics: dry or with light mineral oil.

Forming Operations

For forming (chiefly drawing) operations, there are a few considerations which are generally helpful:

> There is a direct relationship between the yield strength of the stock and the pressure the lubricant has to withstand. Highest pressures are needed for stainless steel, next is high carbon steel, then low carbon steel, brass, aluminum, and finally copper.
>
> Heavier lubricant is needed for tool members made from low carbon steel and cast iron, then for hardened, ground, and polished tool-steel die components.

Greater stock thicknesses require heavier lubricants.

Greater severity of draw needs a heavier lubricant.

Chart 10-5 lists the types of lubricants generally recommended for forming of the most common metals. (*Note:* The list only provides suggestions for the start of trials and experiments to determine the most adequate lubricant for each case.)

Chart 10-5: Lubricants for Forming (Drawing) Operations

Steel: For light cases of shallow drawing, large drawing radii, and comparatively thin stock: mineral oil, water solution of soap, soluble oil (emulsion), or vanishing oil.

For severe cases: heavy mineral oil (alone), drawing waxes, or phosphate coating.

For extreme pressures: chlorinated oils with the addition of chalk, graphite, talc, mica, borax, sulfur, etc.

Note: For facilitating the drawing of steel sheets, the blanks are sometimes coated with a light film of copper before the operation. This is best done by a short immersion of the blanks (cleaned and dried) in a saturated solution of copper sulfate.

Copper and brass: The best lubricant is soap melted in hot water. The more severe the operation, the denser must be the solution. For very severe conditions, use fatty oils or fats mixed with mineral oils.

Aluminum: Light mineral oil alone or mixed with a little fatty oil and diluted with kerosene; mutton tallow and paraffin in equal proportions (for thick sheet) or 30:70 (for very severe draws); lard oil; kerosene with vaseline; graphite melted in kerosene or in lard oil; or water-soluble drawing compounds.

Zinc: The same lubricants are used as for copper and brass, except that the mixtures must be applied hot. In addition, warm kerosene, tallow, or soluble vegetable oils.

Magnesium: For cold forming, any of the usual drawing compounds (as for steel) are satisfactory; for hot drawing, colloidal graphite suspended in a suitable carrier (e.g., low-boiling-point naphtha or alcohol).

Lubricant Application

Methods of applying the lubricant depend on the type of lubricant, its physical and chemical properties (consistency, viscosity, etc.), characteristics of stock, the die, the press, production rate, etc. The lubricant can be:

a. Applied manually by dipping (immersing) the part in lubricant, applying lubricant with a brush or swab, or wiping the stock with a sponge or glove or a modified paint roller.

b. Applied automatically by dripping, roller coating, spraying, flooding, or tumbling.

c. Precoated on the stock as purchased.

Manual application has too many drawbacks: it increases labor costs, decreases production rates, wastes too much lubricant, irregular application causes many rejects, increases safety hazards, etc. Therefore, automatic application methods always should be preferred. Among these there is a new system which deserves mention. It is distributed by the Equipment Division of Franklin Oil Corp. (Ohio) (see Figs. 10-30 and 10-31). Essentially, it is a multispray lubricator in which each pump works independently and individually. Stock for sophisticated and complicated multistation progressive dies and transfer dies may be lubricated properly even within the interior of the dies. The spray pattern is not necessarily round; it may be rectangular or square, and gives complete coverage of the areas to be controlled.

For most cases of shell drawing, the lubricant, either by manual or by automatic methods, should be applied to *both* faces of the blank. It is of the utmost importance that the lubricant should be distributed uniformly on the whole face of the blank, especially in the area which is to be transformed into the vertical walls of the drawn shell. Otherwise, there is great danger of distorted shells of nonuniform height.

Oilers and Wipers

For the automatic cleaning and simultaneous lubrication of strips and coils it is convenient to provide some commercially available oilers and/or wipers. Oilers are essentially boxes containing two lubricant-saturated felt-coated rolls, between which the stock is passed and coated with lubricant on both sides. The excess lubricant is removed automatically and returned to the reservoir. With wipers, the stock merely passes between two oil-soaked felt pads which wipe and clean the stock. In oiler–wipers both operations are combined. In both cases the stock emerges from the lubricating device, clean of dirt and foreign matter and, consequently, die life is increased.

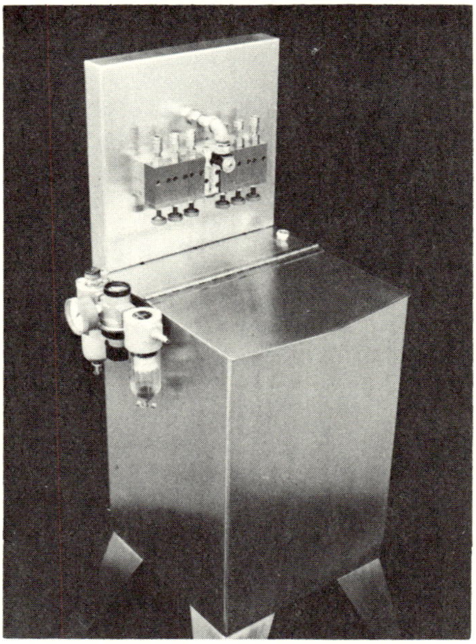

Fig. 10-30. Application of Multi-Spray Lubricator. (*Courtesy of Franklin Oil Corp., Ohio*)

Fig. 10-31. Multi-Spray Lubricator manifold. (*Courtesy of Franklin Oil Corp., Ohio*)

The oiler or wiper may be mounted conveniently on the press or on the die or may be combined with an automatic feeding device and/or stock straightener.

Removing Lubricants

When, in exceptional cases, a lubricant removal operation becomes necessary, consider the time factor; the sooner the lubricant is removed, the easier the operation becomes.

There are no general purpose cleaners or lubricant removers. For each type of lubricant, you must use the correct remover. Some pertinent suggestions are:

1. Solvent cleaning for simple oil-type compounds.
2. Alkaline cleaning for the more-sophisticated lubricant types.
3. Vapor degreasing is sometimes very efficient.

Storing Lubricants

Although it should be self-evident, it is sometimes forgotten that oil containers must be kept closed. Contamination by dust, dirt, or foreign matter will cause problems and expense.

CHAPTER ELEVEN

Ejecting Stampings and Scrap

The ejection or unloading of the stampings and scrap from a die bears a direct relationship to production efficiency. In fact, by means of proper ejection methods, the inactive period of the operational cycle may be shortened drastically and thus the production rate increased considerably.

The ejection is effected in two consecutive steps: first, the workpiece and scrap must be separated from the corresponding tool member and then ejected from the die. The approach to this problem is somewhat different in the case of cutting dies than for forming dies.

Chart 11-1 gives an outline of the different ejection methods used in press tools. The corresponding details follow.

Ejection in First Operation Cutting Tools

In first operation cutting tools the stripping is effected always automatically. The slugs and stampings usually separate themselves from the punches without difficulty and without any intervention by the press operator. However, in certain cases, it is necessary to provide an additional means for the separation such as suction breakers or specially shaped punches. (See Chapter 15.)

The scrap (strip or coil skeleton) is removed from the tool manually or automatically, according to which feeding method is employed.

The removal of the stampings and the punchings should always be automatic and precise, preferably by gravity, i.e., dropped through the corresponding openings in the die plate, die shoe, bolster plate, and press table. (This is the safest and most efficient method of dealing with stampings and slugs. At the same time it ensures the highest degree of productivity. Any other method—such as air jets or mechanical knock-outs—slow down production and are liable to cause problems.) If one of the above-mentioned openings is obstructed or too small, then the stampings and punchings must be removed periodically by the operator. Figure 11-1 shows an arrangement where the die is mounted on a high bolster and the blanks fall into a container. In Fig. 11-2 the container is located in the die shoe.

The arrangement of either Fig. 11-1 or Fig. 11-2 presents a drawback, because it slows down the operation, and, in addition, there is the danger that the operator may forget to take out the workpieces in time and then they may jam the whole operation.

To avoid trouble, in such cases, it is recommended that one of following methods be used according to the characteristics of each case:

a. Tilting the press body so that the stampings, slugs, and scrap pieces slide off the press table automatically into properly located tote boxes.

b. Providing some inclined chutes, either incorporated in the bolster or die shoe (Fig. 11-3) or as separate items. Chutes are always homemade, usually by the die setter.

c. Small low-profile belt or magnetic conveyors, which move in a cavity of the die shoe, or the die may be mounted on parallels (Fig. 11-4)

d. Providing some slides that are actuated and properly timed automatically by the press or by an air cylinder push the blank out of the underside of the die (Fig. 11-5).

e. Sometimes air blasts or vacuum arrangements are used (see corresponding item below).

Note: Methods a and b deflect the motion of the parts, slugs, and scrap. Methods c–e consist of actually removing the material by positive action.

Compound dies work somewhat differently from the other types of cutting dies. With compound dies the finished blank is usually not immediately removed automatically; it either remains on the punch or is carried up into the die plate. In the first case it is often enough to tilt the press so that gravity takes care of the blank ejection, but sometimes it must be aided by a properly timed air blast. In the second case, it must be ejected positively by means of a crossbar or cam-actuated mechanical ejector. This takes time—more than the ejection of blanks through the die plate, as happens with standard blanking—so compound dies are slower and their production rate is lower.

Chart 11-1: Ejection Methods

	Separation	Ejection
First operation cutting dies	1. Strip from punch: always automatic (strippers)	1. Strip skeleton: depends on feeding methods and means: a. manual b. automatic c. scrap cutters
	2. Punchings and stampings from punch: a. usually automatic b. sometimes suction breakers	2. Punchings and stampings: a. usually automatic gravity (direct or indirect, by tilting the press); sometimes conveyors, chutes, air b. sometimes manual c. inverted dies: as with second operation dies d. stacking

Chart 11-1: Ejection Methods (Continued)

	Separation	Ejection
Compound dies	See first operation cutting dies	1. Strip skeleton: see above 2. Punchings: see above 3. Stampings: positive, either air blow or mechanical
Second operation cutting dies	1. Stamping from punch always automatic (strippers) 2. Punchings from punch: a. usually automatic b. sometimes suction breakers	1. Punchings: a. usually gravity (direct or indirect) b. sometimes; conveyors c. sometimes: manual 2. Stampings: depends on feeding method: a. manual: direct (with gripping devices), indirect (pushing slides, pushing rods, pushing plates reciprocating slides, movable dies, dial feeds) b. automatic: gravity (press tilting), air (jet) blow, combined press inclination and air jet, mechanical kick outs, arms, hands c. scrap in trimming and shaving operations: see stampings
Forming dies	Stampings from male and female stamps: a. manual b. automatic: strippers, separators, lifting devices, suction breakers, push-outs, kick-outs (springs, rubber pads, mechanical devices, pneumatic, hydraulic, magnetic devices)	Similar to evacuation of stampings in second operation cutting dies

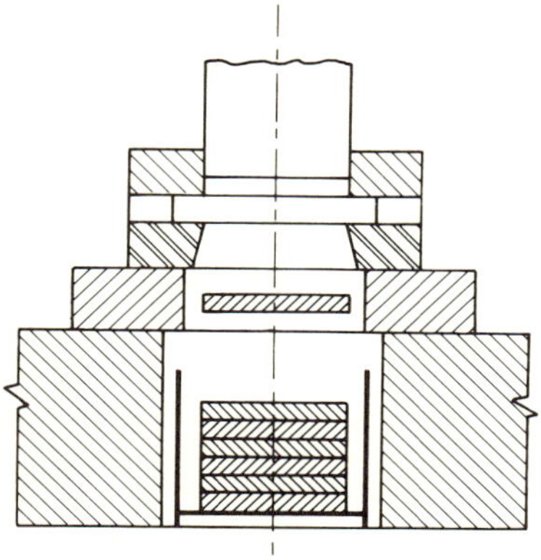

Fig. 11-1. Blanks unloaded intermittently from under the die.

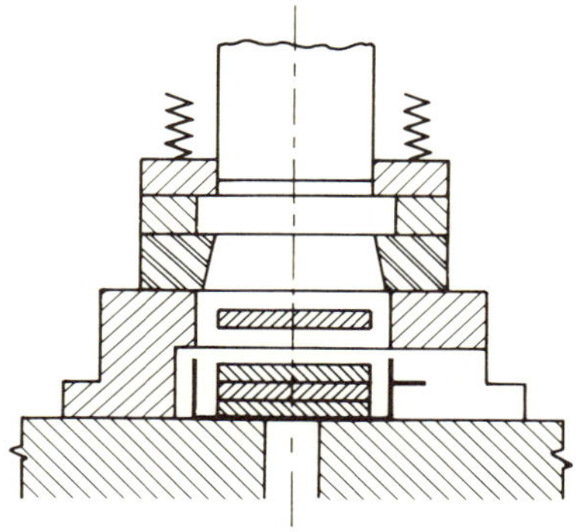

Fig. 11-2. Unloader tray in die shoe.

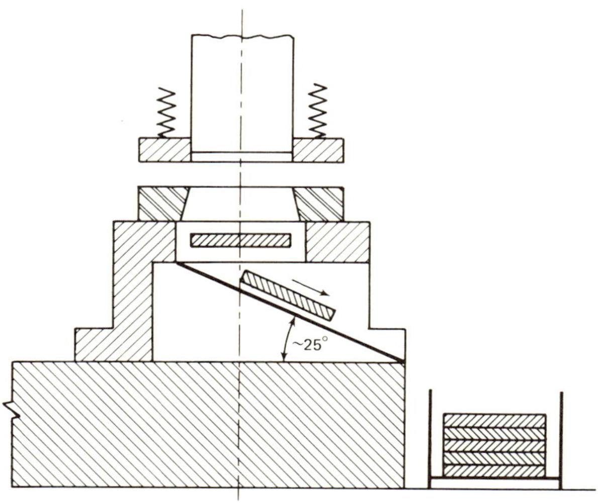

Fig. 11-3. Unloading chute in die shoe.

Ejection in Second Operation Cutting Tools

Here the separation of the stampings from the punch is always automatic and performed by the stripper, analogous to the strip separation in first operation dies. The separation of the slugs from the punch is identical as in first operation dies; usually it is automatic, although sometimes suction breakers or some other auxiliary is needed.

The removal of the slugs is the same as with first operation tools.

The removal of the parts depends chiefly on the feeding method, the die design, the characteristics of the parts, etc., and may be effected manually or automatically.

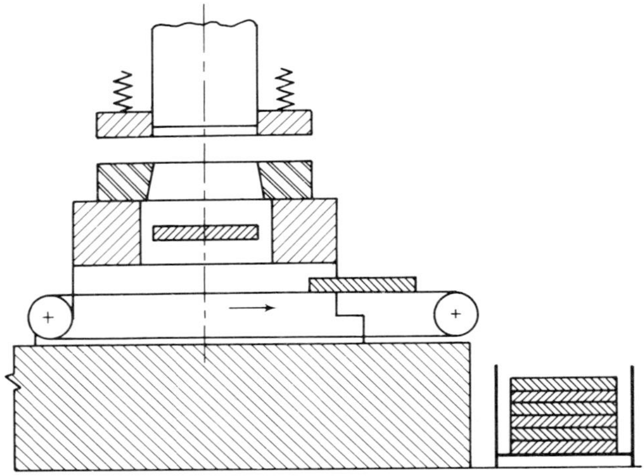

Fig. 11-4. Low-profile unloading conveyor.

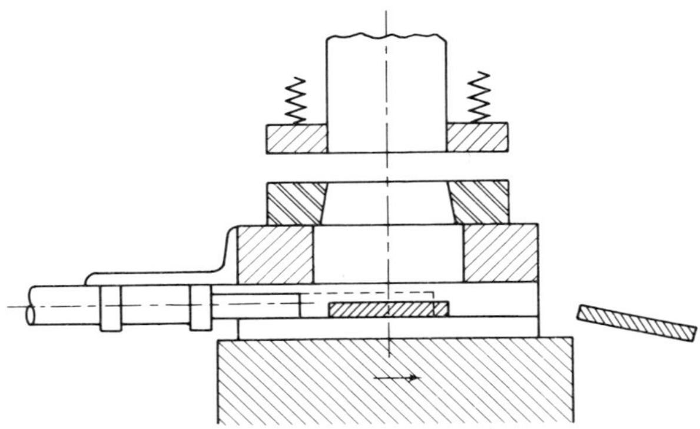

Fig. 11-5. Pneumatic slide pushes the blank out of the die.

Direct manual removal of parts should be absolutely avoided because of safety hazards and of decreased production efficiency. Manual removal of parts should be effected either with special *gripping devices* (see Chapter 9) or by some *indirectly* acting means, such as pushing slides, pushing rods, or pushing plates. These devices literally push the parts out from the die. Sometimes the indirect manual feeding device is employed also for the indirect manual removal of the parts, e.g., reciprocating slides, movable dies, or dial feeds.

Of course, in all cases it is preferable to employ some type of *automatic* ejector, both for greater safety and for better production efficiency. By utilizing simple, inexpensive, ingenious tool members and/or devices the production rate may be increased 100% or even more. An additional timesaving advantage of these devices consists of the indirect shortening of the feeding time. In fact, if the operator is occupied *only* with the feeding, the ejection being automatic, as soon as he or she starts the ram (finished the feeding operation and tripped the press), then, while the workpiece is being processed and ejected automatically, he or she can be ready with the next workpiece to send it or to put it into the die or feeding device quickly, without having to wait for the press to stop.

For the automatic removal of parts the following methods are currently used.

a. By tilting the press, the stampings, freed from the pierced blanks, may slide out of the tool. For light jobs, especially on continually running high-speed presses, the inclination of the press alone is very risky, because it is unreliable; the product must be assisted for safe ejection from the press. Some auxiliary help must be provided; often it is an air blast that helps to accelerate and to guarantee correct ejection.

b. Compressed air is also used alone with both vertical and inclined presses, and usually for small parts. The short, powerful blast must be correctly directed against the workpiece, usually off-center, and properly timed with the ram movement. This is obtained with the aid of valves operated through cams mounted on the main crankshaft of the press flywheel. Sometimes, especially for comparatively large stampings and/or heavy stock, multiple air jets are needed. A usual fringe benefit of air blast ejection is the fact that the air jet simultaneously removes from the die any debris or scrap pieces such as slivers on slugs.

It is good practice not to mount the air nozzle on the press table (Fig. 11-6a) but directly and permanently on the die shoe (Fig. 11-6b). With the latter arrangement, the setting up of the die is much quicker, and its working is much safer and much more effective.

It should be noted that air ejection may be occasionally objectionable because of noise pollution.

c. Mechanical ejectors (knock-outs, kickers, etc.) come in many types and designs: chiefly pins or plungers actuated positively by means of crossbars, cams, springs, hydraulic or air cylinders, sky-hooks, lifters, etc. See a typical example in Fig. 11-7.

The use of springs in press tools in general, and especially for direct ejection purposes, is a rather controversial subject. While springs offer the advantages of being quite inexpensive, simple, and self-contained units, requiring no coupling to the press crankshaft or other member for actuation, they are not always reliable. For very light jobs where there is little or no danger of parts tending to stick or cling in die cavities, springs usually work satisfactorily. But in the case of heavy work, they must be supplemented by some mechanical, positively acting arrangement.

d. Sometimes, as mentioned before, the automatic feeding device, itself, performs

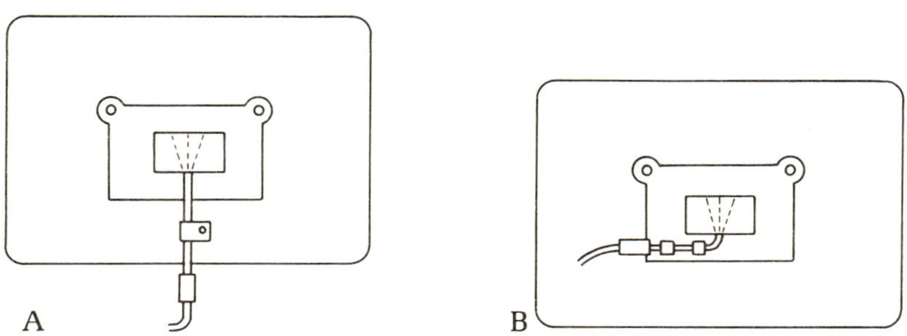

Fig. 11-6. The air nozzle should be mounted on the die shoe and not on the press table: (a) mounted on press table; (b) mounted on die shoe.

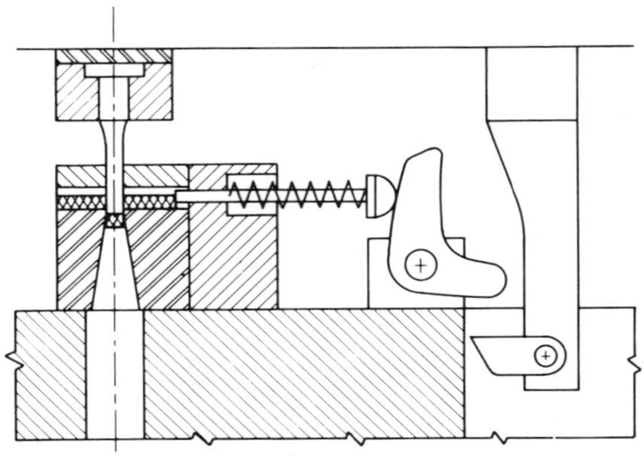

Fig. 11-7. Automatic mechanical ejector for second operation dies.

the automatic ejection of the finished workpieces (e.g., reciprocating slides or movable dies). In dial feeding systems, the automatic ejection of the finished stampings is practically always incorporated at the exit side. It is usually performed with mechanical kick-outs, sometimes aided with air blow-off.

e. Large stampings, e.g., automotive parts, often are removed with mechanical, hydraulic or pneumatic extractors, i.e., gripping fingers, arms, jaws, or hooks, and then often automatically transported with conveyors.

f. In the case of delicate workpieces that cannot be subjected to impacts, such as frail formed stampings or thin-walled drawn shells, they are only separated from punch and/or die and must be removed by hand. However, such a method is unsafe and slow and should be adopted only if and when unavoidable. Therefore, it is better to provide suitable chutes which will guide the ejected parts smoothly in such way as to avoid harming them. In other cases, the parts may be blown gently against curtains which cushion and dampen the ejecting impact and prevent damaging or distorting the stampings.

Scrap Removal

There are some cases—trimming of formed stampings, unit blanking of large flat blanks, or shaving—where scrap material remains on the die plate after the press operation. This scrap must be removed, preferably automatically, by any one of the above-specified methods for part ejection. Often the scrap is "chopped" into small pieces by means of incorporated chisellike cutters or scrap-cutter shears, in order to facilitate its removal and handling.

In other cases of automatic removal of large quantities of steel scrap or stampings, special conveyors are being used successfully. Figure 11-8 illustrates the actual application of a special conveyors with permanent magnets, made by Eriez Magnetics, that is used for scrap removal in metal stamping shops.

In *forming dies* the two steps of ejection—stripping and removal—are not always as clearly defined as in cutting dies. Often there is no stripping at all; in other cases,

Fig. 11-8. Scrap removing magnetic conveyor. (*Courtesy of Eriez Magnetics*)

the two steps are performed simultaneously using the same device. Therefore, the stripping is rarely effected manually, but is mostly automatic. For the latter purpose, if necessary, some adequate device is provided, according to the type of operation, size, shape, and contour of the stampings. There are different types of strippers, separators, suction breakers, lifting devices, knock-outs, and their actuation may be effected by means of rubber pads, springs, air cushions, or mechanical, pneumatic, hydraulic, electric, and magnetic devices.

Here follow a few practical hints for getting easier and quicker part separation:

a. Drawing punches must be perfectly finished in order to permit the greatest ease of stripping the drawn shells from them. For this purpose they should be vertically ground, lapped, and polished, in order to avoid the formation of circular ridges, which are unavoidable in the case of "lathe polishing."

b. In addition, drawing punches should be slightly tapered downward, in order to facilitate the stripping of the drawn shells. Usually a very slight taper is sufficient (Fig. 11-9).

The actual removing of formed stampings is performed basically in the same way and with the same means as blanks processed in second operation cutting dies (see corresponding item above). Of course, whenever possible, absolute preference should be given to automatic ejection either through the press bed bottom (e.g., flangeless shells or boxes) or by gravity (inclined press).

Here are a few practical suggestions for part removal:

a. Ejectors for small formed workpieces are sometimes very difficult to place in tools which are usually of very small dimensions. In such cases, it is sometimes convenient to permanently magnetize the punch so that the stampings remain adhered to it and are lifted automatically when the press ram ascends (Fig. 11-10). At open tool position, the finished workpieces are easily removed from the punch or are separated from it by compressed air or by mechanical means. Of course, this method is possible only if the stampings are made from steel.

b. A positive ejector design (which may be combined with an inclined press and/or an air jet) is shown in Fig. 11-11. This method offers the possibility of providing higher ejecting forces than springs so it is especially suitable for heavy stock, which has a greater tendency to cling to die cavities.

c. As ejectors are very small drawn shells in a die, where the usual spring-actuated or positive-action ejectors cannot be employed for lack of space, small pieces of rubber, cut from a common eraser, may sometimes be used successfully (Fig. 11-12). This method is recommended for shallow workpieces with ample outside taper.

d. For difficult cases where the drawn (flanged) cup must be extracted from the die by means of the punch itself, the punch is made with a knurled end. The bottom part of the punch for a short length (0.2–0.3 inch) is made somewhat smaller in diameter—approximately 0.003 inch—than the specified size. The knurling then increases the punch diameter to the exact size (Fig. 11-13).

Stacking of Stampings

Stamped blanks, as they come out of the dies, are usually gathered in bins or tote boxes for further handling and processing. Either directly by gravity or indirectly by the aid of small conveyors or by hand, they are thrown into the tote boxes quite irregularly. This is all right for most cases; however, for certain kinds of stampings and/or conditions, it may represent a serious drawback particularly where the stampings have to be used in subsequent secondary operations. The solution for such cases is stacking, which consists of depositing the stampings in vertical, horizontal, or inclined stacking devices.

Advantages of Stacking

First of all, there is an increase of production rate. In fact, the irregular way stampings ordinarily lay in a bin means loss of precious time if, for the next operation (press working, inspection, assembly), the parts must be fed or presented or delivered in a well-determined orientation (part shape presented in a given direction, burr side up or down, etc.)

The most typical cases are motor laminations for stators and rotors, transformer laminations, and other similar kinds of stampings that must be assembled into riveted, laminated assemblies.

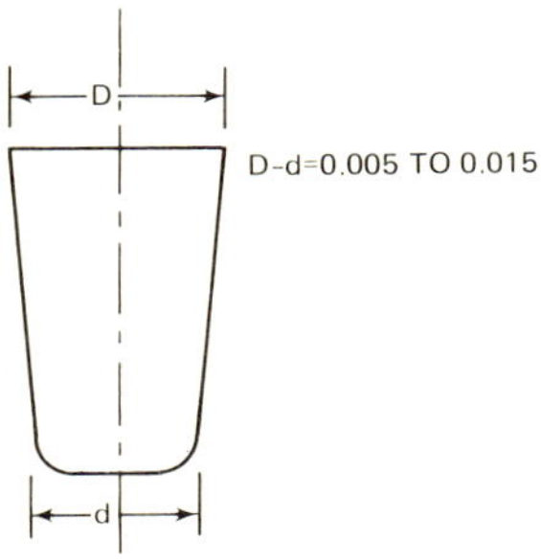

Fig. 11-9. Slight taper on drawing punch facilitates ejection.

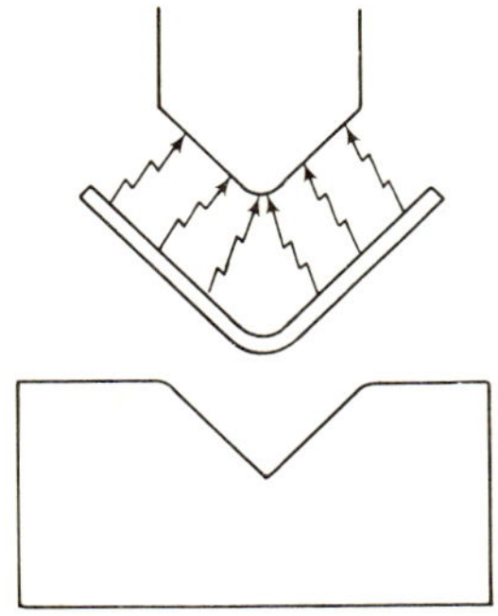

Fig. 11-10. Magnetized forming punch lifts formed stamping for easy manual removal.

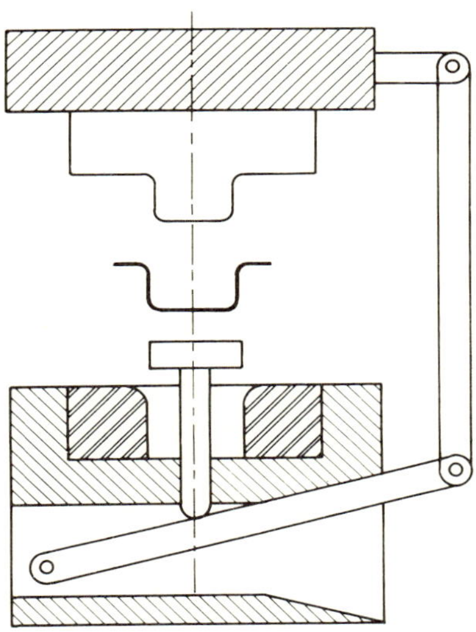

Fig. 11-11. Positive mechanical ejector for formed stampings.

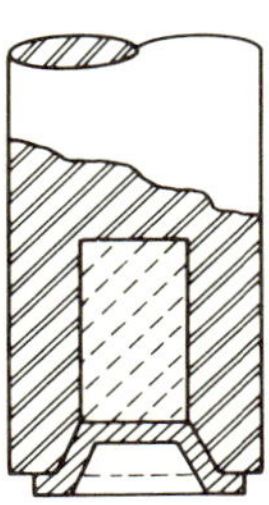

Fig. 11-12. Rubber pad ejector for small formed parts.

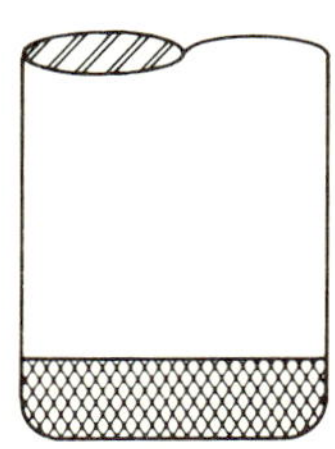

Fig. 11-13. Knurled drawing punch helps to lift drawn shells from die.

The role and importance of stamping orientation is usually overlooked; it is not given sufficient consideration, in spite of its notable effect on production rate, as will be seen.

If we eliminate the need for the separate orientation of each part for the next operation, we get more production than by greatly increasing the speed of the corresponding press. A numerical example illustrates the correctness of this statement. Let us take a low value of 1 second for the part orientation. A low-speed press of about 120 spm requires for one complete press stroke 60 ÷ 120 = 0.5 second. If this press is changed for another (hypothetical) one having a speed of 1200 spm, this latter requires (60 ÷1200 = 0.05 second for one press-stroke. As seen, the time saving realized by the tenfold increase of press speed (0.5 − 0.05 = 0.45 second) is less than one-half that resulting from the correct *automatic* locating of the workpiece. Consequently, the unloading method of the first operation must be carefully studied, to provide the best method of gathering the workpieces and locating them automatically in bins or magazines in the same position, i.e., by stacking.

While the chief advantage sought with stacking devices is the increase in productivity, i.e., savings of costs, time, labor, there are several additional fringe benefits to be gained as specified below.

There is a better protection of the surface finish of delicate sheet metal components. Sheet metal with a high polish is easily damaged (scratched, nicked by burrs, punchings, etc.); such conditions may convert the otherwise acceptable stampings into rejects. In severe cases of very rough handling, the stampings may be distorted by abuse. In all these cases, stacking provides protection for the workpieces.

There is a further advantage of stacking when the stampings are not needed immediately for the subsequent processing, as is often the case. The handling, transportation, storage, and inspection of the stampings is made much easier and simpler and therefore less expensive. In many cases stacking is employed only for such reasons, e.g., when the stampings are already finished in the last operation and there is to be no further press processing.

Furthermore, stacking of stampings increases safety to the people who handle them, because there is less of a chance for hand and finger injuries.

Finally, stacking of components facilitates checking and counting.

Design of Stacking Devices

The design of stacking devices depends on the stacking method; the form, shape, and size of the components; the design and construction of the press tools; and the characteristics of the press.

Classification of Stacking Methods

Although there are many varieties of stacking methods, these may be reduced to a few basic types:

1. the stampings are pushed directly into magazines attached to the press, just below the die;
2. the stampings fall freely into magazines located below the press, underneath the die;

3. the stampings are conveyed through inclined chutes to loose magazines located at convenient distances;

4. the stampings are guided through curved chutes, with open unturned ends; the blanks are then removed by hand periodically.

In any method, the actual guiding may be effected either by the outer contour or by some opening in the stampings or both. The guides should have a rather loose fit to the stamping details.

Construction of Stacking Devices

Magazines and chutes must be designed as interchangeable units in such way that it is possible to produce similar stampings on different presses, or a given press should be able to handle a certain variety of workpieces.

Magazines and chutes should be large enough so that their capacity is sufficient for a certain, long time interval until they are filled—30 minutes to 1 hour is considered as a good average.

Often the stacking device is made in such form, shape, and size that it can be employed *directly* as an interchangeable, replaceable magazine for the feeding device of the next, second operation tool.

The magazines and chutes may fully surround the stampings or only partially, controlling the strategically important portions of the blank or hole contour; in some cases they are simple skeletons built from rods or pieces of tubings and plates.

Magazines Attached to the Press

Figure 11-14 shows a simple example of direct stacking, employed for round, square, rectangular, or similarly shaped simple stampings. The magazine is simply slid into place with suitable guides and indexing stops provided; in other designs the magazine is hung in its place below the bolster plate and held there by means of flat springs or other suitable devices (Fig. 11-15).

In operation, as the blanks are stamped, they are pushed down one after the other into the magazine against an adjustable, spring-actuated friction drag (made from rubber, leather, etc). This arrangement (providing an adjustable, small back pressure) is necessary because of the fact that the magazine has only a loose fit with the stampings and if these were to fall down, they would turn around in the magazine and probably jam it. Another method for furnishing the back pressure consists of employing calibrated compression springs as shown schematically in Fig. 11-16. Finally, the back pressure may be furnished also by simple, light counterweights, which balance the drag plate (Fig. 11-17).

When the magazine is full, the press must be stopped automatically and an optical and an acoustical signal alerts the attendant. Then a thin sheet metal piece is slid in at the underside of the tool or bolster plate in order to prevent those stampings which are in the die plate plus die shoe plus bolster plate from falling down. The full magazine is then removed and is replaced by an empty one.

Sometimes, when there is not sufficient space underneath the press, the magazine is located horizontally and the stampings are conveyed into it by means of short curved chutes (Fig. 11-18).

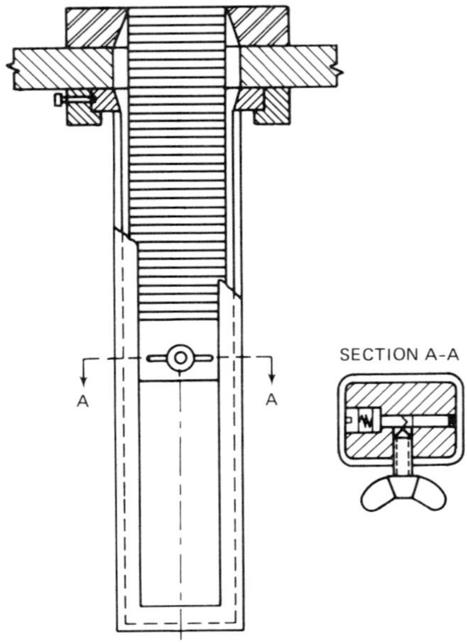

Fig. 11-14. Direct stacking with slide positioning.

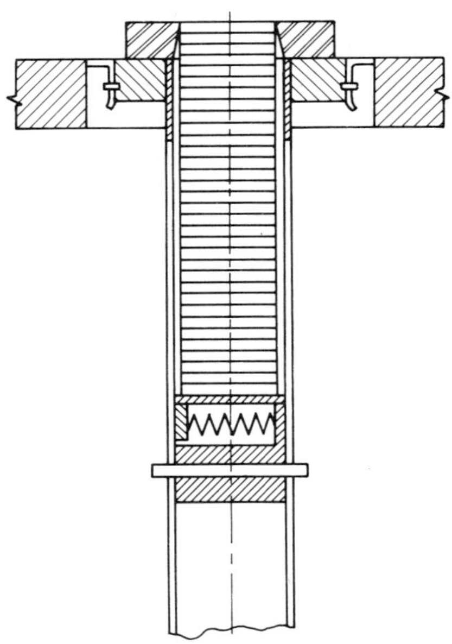

Fig. 11-15. Direct stacking with hanged positioning.

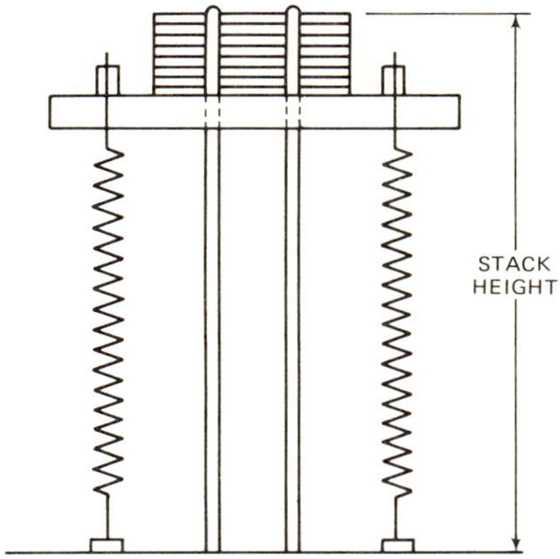

Fig. 11-16. Back pressure through compression springs.

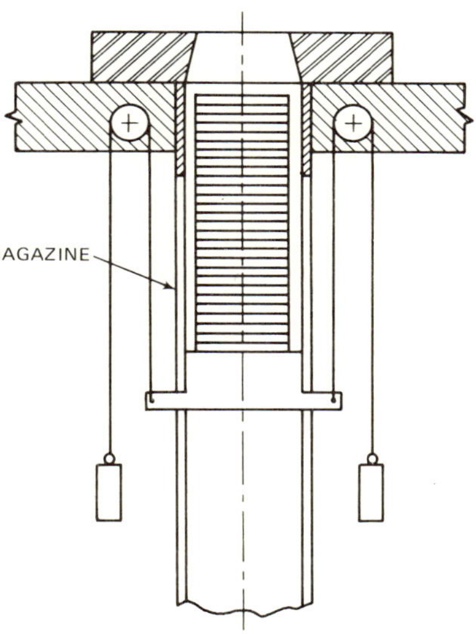

Fig. 11-17. Back pressure through counterweights.

Loose Magazines

In certain cases, e.g., when the stacking is effected by some punched hole, it is possible to eliminate the back-pressure arrangements: the stampings are simply permitted to fall on a vertical rod, which constitutes the core of the stacking device. Figure 11-19 illustrates this principle. The device in question consists simply of a guided slide carrying a mandrel. It must be watched by the press-attendant in order to stop the press when the device is full and replace it with an empty one. In these cases the press may be stopped automatically; e.g., after a predetermined number of press strokes.

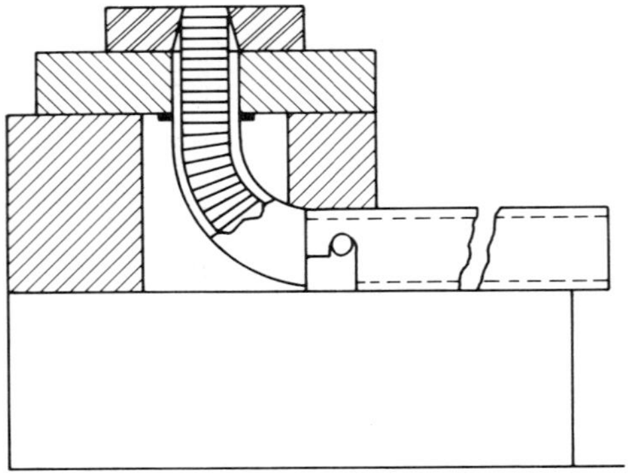

Fig. 11-18. Curved chute for conveying the stampings.

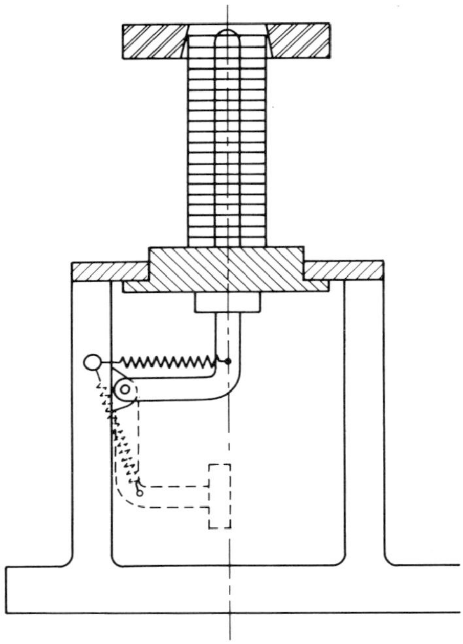

Fig. 11-19. Loose magazine with central mandrel alignment.

Figure 11-20 shows an analogous case where two different stampings are produced in one common progressive die, which reuses the slugs to improve the stock utilization factor. Two, correctly located stacking devices collect the stampings, guiding them by means of suitable mandrels.

Selectivity of Stacking Devices

A stacking device must be, logically, selective. This means that in the case of progressive and/or multiple dies, each workpiece type must be stacked separately from the other(s) and from the scrap (punchings, etc.). Figure 11-21 shows a simple,

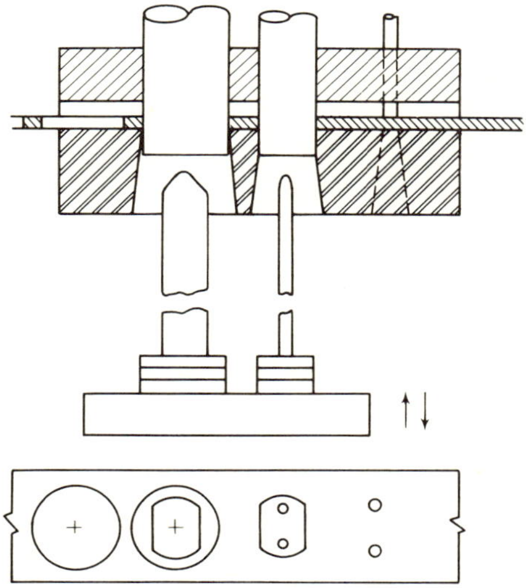

Fig. 11-20. Loose magazines for selective stacking.

typical example. In a progressive die two different sizes of washers are produced together (another case of scrap utilization). Both sizes of stampings slide down a common chute which has a suitable opening where the smaller washers get through, but the larger ones do not. In this way, the stampings are caught at separate stations by their respective mandrel-type stacking device. Figure 11-20 also shows selective arrangement since the two stampings are collected separately by the two mandrels.

Conveying of Stampings to Magazines

This method is essentially a modification of that shown in Fig. 11-20. It is used whenever the stampings cannot be dropped vertically down, for example, in compound and inverted dies, for stampings that are too large, and for a bolster plate without clearance holes that are sufficiently large or that are completely absent. In all these cases the stampings are conveyed to the magazine. This is effected either by gravity (simple inclined chutes) or by powered conveyors.

In Fig. 11-22 the stampings slide down the chute of the inclined press and are stacked on a mandrel by means of their central holes The magazine of Fig. 11-23 stacks the stampings guided by their outer contour. The tension spring must be so calculated that the weight of one stamping will lower the magazine exactly by the height of the stock. In such a way, the top blank in the magazine always has the same vertical position.

Curved Chutes

It seems that this type of stacking device is the one most widely employed. The reason for this preference is probably due to the fact that its operation is the simplest and its productivity is the highest. In fact, there are no magazines to change and, therefore, the press is not stopped periodically. The production is continuous, and

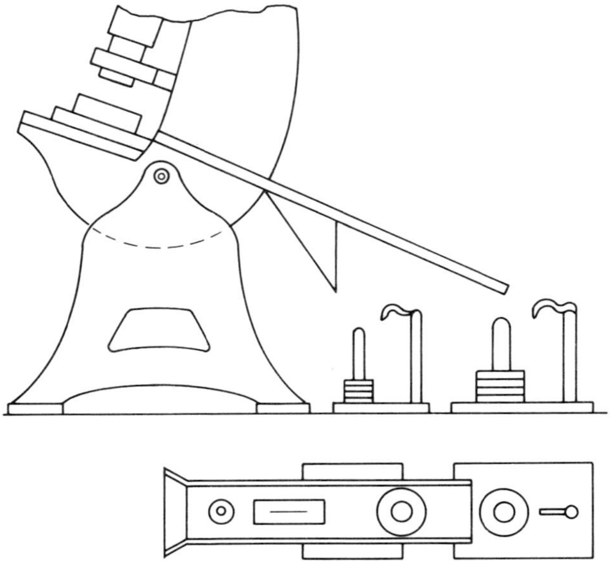

Fig. 11-21. Selective stacking through special chutes.

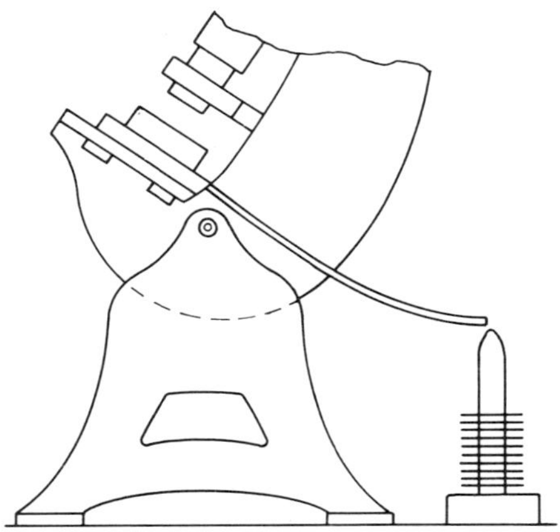

Fig. 11-22. Conveying stampings to a loose mandrel.

there is no lost time. The unloading is effected while the press is running. Indeed, the only attention that must be paid to the press consists of removing periodically, at approximately predetermined intervals of time, the finished stampings from the chute. The chutes are attached to the underside of the press table or bolster plate.

Such chutes are very simple devices. As shown schematically in Fig. 11-24 they are usually made from a skeleton tubular or solid rod construction, with ample clearance, thus providing a practically frictionless guide channel. The guide rails are obviously so positioned as to suit the blank contour and are held in place by retaining

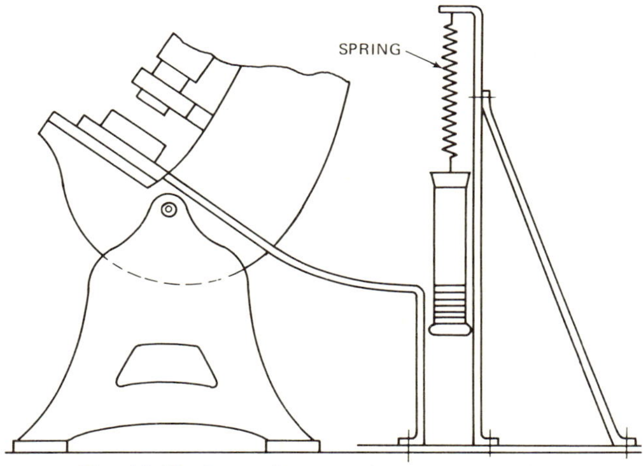

Fig. 11-23. Conveying stampings to a magazine.

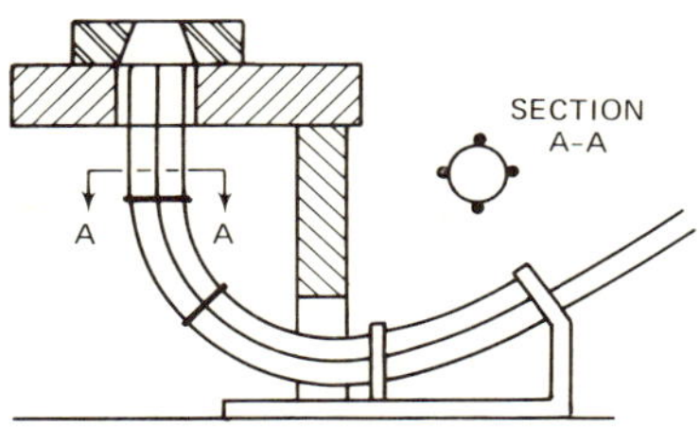

Fig. 11-24. Curved chute.

rings. The chute is bent to a large radius in order to provide smooth travel for the stampings. It is practically always made in short lengths suitably connected together with flanges for ease of installation, inspection, service, and maintenance. Figure 11-25 illustrates another example where two kinds of stampings ("E"s and "I"s of transformer laminations) are stacked separately at two sides of the press.

In a typical design of a curved chute there is a short vertical tubing section just below the press bed, connected with curved portions and finished with a straight, slightly upward inclined section; sometimes there is an additional completely horizontal exit section. From the last open section, the workpieces are removed periodically.

Also with curved chutes, since the stampings are ejected from the die positively, it is necessary to build up a proper back pressure in order to prevent jamming of the chute (see corresponding item above). The back pressure is provided by the weight of the stampings which fill up the chute in the upward turned sections. At the start, a proper adjustable friction drag must be provided, as in the arrangement with vertical magazines.

The problem of back pressure does not always arise. In fact, sometimes the stacking device is made with a horizontal mechanism. Figure 11-26 shows a case where the guided stampings are conveyed by gravity in such a way that they become properly stacked in a horizontally oriented stacking device.

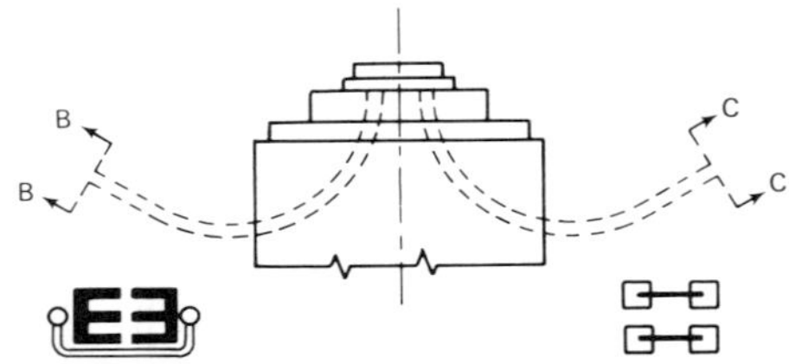

Fig. 11-25. Selective stacking with curved chutes.

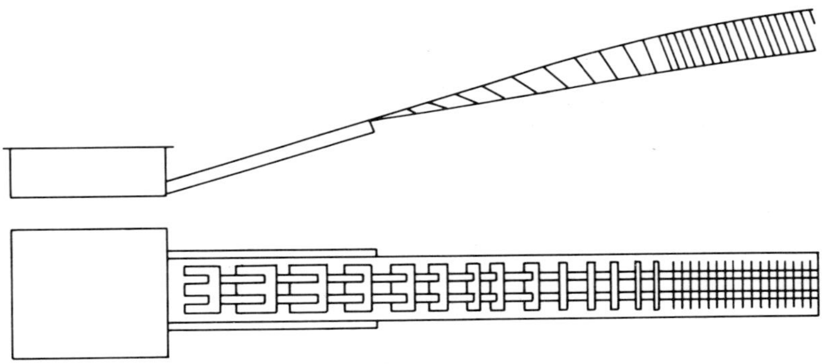

Fig. 11-26. Stacking without back pressure.

Indexing Stacking Devices

In some cases the productivity of the stacking magazines may be improved by eliminating the need to periodically stop the press in order to change the magazines. For example, the arrangement of Fig. 11-22 may be modified, accordingly, in the following way: A hydraulically or electrically or pneumatically actuated dial table carries a certain number of removable mandrels. One of them is always correctly positioned to receive the conveyed stampings. As soon as the first mandrel is full, i.e., after a preestablished number of press strokes, the dial table is turned automatically to the next, empty mandrel, and so on. The press attendant removes the full mandrels and replaces them with empty ones while the press is running without interruption. Thus, there is no loss of time, and the production rate is improved considerably.

Limitations of Stacking Methods

The process of stacking stampings may be performed trouble free and satisfactorily only in case of flat blanks and/or shallow formed parts. Deep formed (bent, shaped, drawn, etc.) stampings are not easily stacked because the space needed in the axial direction would be excessive. Besides the slightest asymmetry of the stampings would impede correct operation of the stacking device.

Automation of Press Operations

In the metal stamping industry, and in industries that use metal stampings as components, many experts think that one of the most efficient methods for improving productivity consists of the application of automation. The concept of automation, basically, may be expressed simply in this way: Replace manual operations by fully automatic, mechanical operations, i.e., eliminate almost totally any direct human intervention in industrial production. The final goal, utopian in most cases, is the manufacturing plant with all its machines synchronized and operating practically without workers, where everything happens automatically.

Theoretically and also practically, automation reduces costs, accelerates production, and provides many other benefits.

The word "automation" may be used loosely as simply synonymous with "mechanization," i.e. transforming a manual operation into a semiautomatic one, or a semiautomatic operation into a fully automatic one. This is not true automation. True automation means not only mechanization of the different operations, but also their blending into a fully automatic interoperational production line including every stage of manufacture.

According to experts, the benefits of a totally automated factory are reserved for a very few industries where enormous lot sizes are required of the same product (a typical case being cans for canned products). For the average shop or plant, total automation is out of the question; it is not realistic. This is true for the metal stamping shop.

However, it would be irrational and wrong to renounce automation, simply because it is not suitable for complete adoption in a given plant or industry. *Partial* automation often is realistic and one can and should try to do something in this respect.

To automate some operations—provided it is justifiable both economically and technically—generally does not present major problems, and the advantages achieved are several, interesting, and important: increased production rate; decreased waste and rejects; elimination or at least a drastic reduction of industrial accident hazards; labor savings; freeing of workers from a monotonous, repetitive activity with a rhythm dictated by a machine.

Productivity is increased principally because the press utilization factor or active press hits increases from 20 to 30% with manual feeding methods, up to 70 to 80% with mechanical feeding, besides being improved by total automating of the evacuation of the parts and scrap and by decrease in rejects due to the more exact location of the blanks.

Partial automation of press operations may vary between application of simple feeding and/or unloading devices and total elimination of material handling between operation points.

It is interesting to note that large investments are not always required to make a plant partially automated. In addition, it is not necessary to set out with a highly ambitious goal. One may perfectly well establish a modest program which is put into effect step by step. And each step can represent an achievement in itself without any absolute necessity to continue with the rest of the program.

The following presents, approximately, a reasonable process of introducing automation in a metal stamping plant:

First operations

1. Substituting calibrated coils for strips sheared from sheets.
2. Automatic or mechanized evacuation of products, slugs, and scrap
3. Fully automatic feeding
4. Eliminating (at least reducing drastically) secondary operations, by means of (a) progressive dies, (b) combination dies, (c) compound dies, (d) transfer dies, (e) transfer presses
5. Automatic counting, stacking, and storing of products
6. Automatically checking the products, and avoiding rejects by stopping the press in case of trouble.

Secondary operations

1. Automatic evacuation of products and scrap (manual, individual, indirect feeding)
2. Manual indirect feeding
3. Semiautomatic feeding
4. Fully automatic feeding
5. Automatic checking, stacking, and storing of products
6. Automatically checking the products, and avoiding rejects by stopping the press in case of trouble

Interconnected production lines (additional operations)

1. Semiautomatic interoperational material handling
2. Fully automatic interoperational material handling

With only the application of automatic feeding and unloading devices, the possibilities of automating are not exhausted. A fertile field is represented by automating arrangements around the working place, specifically: interoperational material handling, scrap handling, and auxiliary operations.

Partly processed parts that must be transported from one workplace to the next should be carried over the shortest possible distance and by some means that eliminates human intervention completely. Inclined planes, roller conveyors, fixed or movable belt conveyors, etc., can be used for this purpose. Such devices can provide material travel throughout the plant with consequent reductions in material handling. If the following presses are equipped with feeding devices, the arrangement is converted into a semiautomatic chain production system. Of course, as always, it is necessary that such arrangements should be justified economically. Usually savings start exceeding investments when over 50,000 parts are produced.

One of the auxiliary operations deserving special mention is semiautomatic or fully automatic lubrication of stock. Everyone connected with metal stamping production should be well aware of the advantages obtained with automatic stock lubrication devices, since they help press operations to procede without trouble and without interruptions. (See Chapter 10.)

Economic Justification of Automation

With mechanization and automation of the production processes of metal stamping shops, certain economic savings must be obtained, or the working conditions must be made substantially easier, or safety must be increased or a combination of these goals must be achieved. If not, the expenses created by the change cannot be justified.

The financial effects of mechanization can be calculated quite easily. This computation must be made anyway before the purchasing of the corresponding automatic devices, because it is the economic gain which principally determines the value of mechanization.

The corresponding calculation is approximate and is based on average values. Essentially it is this: The savings in labor costs and overhead, after the introduction of automation must be higher than the amortized cost of the installations necessary for the automation. Therefore, the following simple equation must be satisfied:

$$M_1 + K_1 \geqq M_2 + K_2 + B$$

where B = cost of devices employed for automation
M_1 = yearly labor expenses, without automation
M_2 = yearly labor expenses, with automation
K_1 = yearly overhead, before automation
K_2 = yearly overhead, after automation

Note: When automation is introduced, overhead increases somewhat: there are more maintenance costs; the setting of dies and devices in the presses requires more time; and, in general, operators must have better training than in the case of conventional production.

The above formula is based on an amortization of one year. If a slower or quicker amortization is contemplated, then the formula must be modified accordingly.

The decision about automation should not be based entirely on economic factors. There are many additional advantages obtained through automation, advantages which cannot be easily expressed in numbers or in dollars and cents. One must take in account such factors as:

a. Shorter delivery dates due to higher production speed (rates). This may sometimes represent the difference between success and failure of a commercial venture (e.g. seasonal articles which are delivered on time or too late).

b. Accelerated material flow; with less unused raw material in a given period of time.

c. Decreased storing space is needed both for raw material and semifinished parts.

d. Decreased safety hazards. Accident prevention conditions improve considerably because press operators do not have to intervene physically in the press operations. (OSHA and insurance benefits.)

Technical Preconditions for Automation

On the other hand, automation has its own limitations due to its exigencies. Remember that an automatic operation is not always necessarily also an economic one.

Before starting to apply automation to a metal stamping plant or shop, many details must be taken care of and many conditions must be satisfied. The main ones of these details are specified here.

1. Dies and their accessories, presses and their attachments, and other pieces of equipment must be in good working and operating condition for rather long periods of time. The smallest failure may have very grave repercussions on production. If just one machine stays inoperative for a considerable period of time, this requires stopping a whole line of presses which form the group—a chain or line. Every die detail that may fail, must be anticipated and taken care of in the design. Take extra care in preventing screws and bolts from becoming loose and springs from breaking. Since these latter components are usually among the first to give trouble, it is recommended that preference be given to rubber or some other kind of springs or to air cushions, instead of steel springs. In general, the necessity of easy and quick maintenance must be provided for in the design and building of the dies.

2. Automation requires some conditions not called for in conventional production, where the press operator or inspector performs certain functions. Thus, tools and installations must be equipped with automatic checking and inspecting devices to ensure: correct feeding of the blanks; their correct location in the dies; removal of the finished parts (and eventually stacking them); scrap removal (especially punched slugs which may create grave problems); and checking the finished part as to whether or not all the specified holes actually exist (e.g., the breakage of a small perforating punch must be detected). All of these and other details must be watched and supervised automatically. Failure of any of these details to function correctly may have disastrous consequences. Therefore, it is imperative that efficient automatic checking be provided which will immediately stop the press in case of irregularities, even the smallest ones.

3. The workpieces must be designed for fully automatic handling. Much closer tolerances are required than those strictly necessary for the function of the parts. Large parts formed from thin gage sheets are usually very weak and deform very easily, making it difficult to maintain a regular, correct, and unhindered material flow. As a matter of fact, at the present time only a quite low percentage of metal stampings would be suitable for automation, without some more or less substantial modifications.

4. The requirements with respect to raw material are stricter. Variations of thickness and of other dimensions, e.g., strip-width, must be within closer limits, uniformity of texture and hardness must be better, and, in general, quality must be higher. If not, there is danger of irregular functioning, with serious consequences.

5. The quality, training, and preparation of the workers must be on a higher level than for conventional production. both white-collar people (die designers, production engineers, industrial engineers, etc.) and blue-collar people (toolmakers, die setters, maintenance personnel, etc.) must be knowledgeable not only in toolmaking, etc., but also in electricity (remote control), electronics, hydraulics, and pneumatics. In general, the workers including the press operators must be better prepared (more technical education and training in specialities) and must have a

higher sense of responsibility than the standard worker. Also, the ratio between white-collar people and blue-collar employees changes; with automation, obviously more technicians and engineers are needed, and fewer low-skilled workers.

6. Take into consideration also the commercial aspect of the matter. Since tooling set-up and die assembly of set-up require much more time and labor than conventional die setting, automation makes sense from an economic standpoint only for the case of very high quantities. Correct automation can mean an enormous increase of production. Since these products must be sold, the sales organization of the company may have to be increased in order to be able to take care of the changed conditions.

7. Last but not least: Can you afford it? Implementation of automation requires quite high investments.

Therefore, before deciding about automation, an exhaustive technical and economical evaluation must be made and the proposed application of automation must be studied carefully in order to determine whether or not every aspect or at least the majority of the important ones warrant the conversion.

CHAPTER TWELVE

Inspection of Stampings

Not too many years ago industrial inspection consisted simply of examining finished parts or products in order to see whether or not they met specifications. In other words, the traditional inspector separated the good parts from the bad ones.

Today, inspection has another, more important role: in cooperation with production, it detects when problems arise, thus helping to prevent—or at least to reduce drastically—spoilage, rejects, waste, and unsatisfactory work.

Inspection in the press shop is continuous: it starts with the incoming raw material, continues with parts and components in process, and finishes with the assembled end products.

Inspection and quality control directly contribute to production efficiency. Inspection is decidedly one of the chief production means for manufacturing the most satisfactory product of the highest quality at the lowest cost.

The time, work, and money spent on inspection is variable, depending chiefly on the accuracy required and the reliability of the workers. In any case, inspection is costly: in some exceptional cases it may cost even more than the actual production. Therefore, it is worthwhile to do everything to improve the efficiency of this activity.

As happens with material handling, inspection increases the cost of products without adding to their value; therefore, management must avoid quality control and inspection systems that are to elaborate, where the expense may exceed the benefits derived from it. Nevertheless, it is an established fact that whenever an efficient inspection department is set up, the savings realized through the prevention of waste are practically always greater than the cost of the department.

Pertinent literature seems to indicate that inspection engineering is concerned predominantly with machined workpieces, die castings, cold-headed parts, etc., to the almost total neglect of metal stampings. Therefore, this chapter is based mostly on personal experience and that of colleagues and friends engaged in metal stamping production.

The approach to inspection of metal stampings is rather different compared to that for machined parts. As will be seen, stampings are produced at such higher rates that it is uneconomical and impractical to inspect them all individually (100% inspection); only spot checks are reasonable. In addition, press tools have rather constant and long-lasting characteristics so that changes in product size and/or shape are not likely to occur suddenly; these problems give early warnings.

This chapter is intended to answer such basic questions as: When to inspect?

Where to inspect? By whom? What to look for? It is also meant to give a short presentation of the means and methods of inspection currently used in press shops.

When to Inspect?

Metal stampings must be inspected (1) at the start of a production run; (2) during the production run; and (3) at the end of the production run.

At the beginning of each run, the first few stampings must be inspected. Only if they pass is regular production allowed to begin.

First piece parts must be inspected after each partial or total die setting. In many shops, the first parts must be checked at the start of each shift. (This is an excellent practice that often prevents serious trouble.)

The stampings produced during the production run must be inspected from time to time. Obviously, as stated before, the older practice of final inspection alone reveals only that the lot is satisfactory or unsatisfactory and, in the latter case, the causes of the loss, but it does not prevent the damage. Only preventive inspection, performed during the production run, can do this. Therefore, inspection must be conducted during the stamping process, periodically, in the form of spot checks, in order to stop the production at the first sign of discrepancies or trouble. If the problems' causes are found out early in the run and eliminated while still small, their harmful consequences are greatly diminished.

The intervals at which the inspector checks each machine vary from 20 minutes to a few hours (30 minutes being the preferred standard for many shops) according to production rate, characteristics of the stampings, part tolerances, reliability of the press operators, etc.

It is excellent practice to inspect the last few stampings after completion of a production run. In many shops, this is a standing rule: no die is removed from the press without checking the last pieces and the corresponding strip skeleton.

Where to Inspect?

Workpieces must be inspected as near as possible to their origin, near in location and near in time.

Near in location means at the press. Any additional handling or transportation complicates the matter, increases the cost, introduces the possibility or even the probability of additional defects and errors which then give an incorrect picture of the press operation.

Near in time means before the problem becomes too serious. As was mentioned before repeatedly, a small defect, if not discovered in time, may develop into a big problem, with very unpleasant and costly consequences.

Only the final inspection is independent of location and time; therefore, it is preferable to perform it in a separate, adequately equipped room.

Inspection Personnel

Inspectors must be highly skilled and trained technicians. They must be well-experienced in the use of standard and special measuring and testing instruments and devices, and they must know all about the parts and products they inspect and the pertinent production operations and processes used.

The chief inspection is usually performed by "flying" (roving, floating, circulating, patrol, line, floor) inspectors who at more or less regular intervals check the characteristics of samples of work at each press in their section against the specifications. By the way, it is a sound practice not to take only the most recent stamping, but also to pick out from the tote box some samples at random.

Since the inspector can visit each press only at long intervals (in some plants there is one inspector for as many as 40 presses), it is the duty of the foremen and of the die setters and even of the press operators to examine the parts more frequently, checking their accuracy by comparison with blueprints, gages, or master samples. If something is wrong, or even suspicious, the press must be stopped at once and a superior notified immediately so that any necessary corrective steps or action may be taken.

It must be emphasized that there is a coresponsibility between inspectors and production people, who should not wait until the inspector comes along to do his checking. With modern high-production presses, quite large quantities of stampings are produced in but a few minutes. If something goes wrong before the next visit of the inspector, much stock, labor, and other expenses may be irretrievably lost, because the stampings will be rejects.

Under normal conditions, there is a certain "hierarchy" in inspection duties, according to different levels of responsibility, training, etc. For example, the operator would be responsible only for errors discernible with the naked eye (missing holes, scratches, nicks, etc.) or by means of simple devices. From the foreman and die setter more detailed inspection would be expected, i.e., they should be able to read blueprints, and to measure and check the stampings with precision tools (vernier calipers, micrometers, etc.) and with gages. Finally, the inspector should be trained to handle sophisticated measuring devices (gage blocks, mechanical and other amplifying comparators, optical comparators, etc.) that may be needed.

Of course, under special conditions—for example, in either very small or big plants or in shops with very rough or very fine products—it may be that the operator must also act as a highly skilled inspector; on the other hand, even the foreman may be required to use only simple measuring devices. In any case, it is chiefly the press operator's responsibility to produce acceptable parts.

In a shop of normal working-force distribution (enough inspectors in proportion to direct labor, particularly where there are fully automatic presses), the inspection activity should be scheduled so that at least one inspector is present in the inspection room all the time.

Unfortunately there are differences of opinion about the authority of inspectors. Theoretically, the inspector's judgment and decision—about what is a reject and what is an acceptable part—is final. However, it happens quite often that for one reason or other the inspector's decision may be overruled by the superintendent, manager, engineering department, etc. When customers then complain and return the questionable parts, usually, nobody remembers who did the overruling. In such circumstances, the inspector should always insist on written orders.

What Is a Reject?

Any product or component must fulfill certain requisites of (a) dimensional accuracy, (b) shape accuracy, (c) finish or appearance, (d) strength, (e) performance, or (f) quality, indispensable to its function or task. *Any* unsatisfactory or substandard detail makes the workpiece a reject.

In actual production, it often happens that some defective parts may be salvaged by reworking. Nevertheless, such salvaging requires time, labor, and money and should be avoided. It may be considered that rejects cost as much as good parts but as a matter of fact, rejects entail additional costs for their identification and separation from the good parts.

Here follow some basic concepts of possible causes for rejects.

Stock: Since it is impossible to produce satisfactory stampings from inadequate raw material, inspection must check all incoming stock. Such inspection must include control of size (thickness, width, length), physical and mechanical properties, surface finish conditions, and, in some special cases, chemical, magnetic, and electrical properties. An Olson test or the equivalent may be used to determine formability properties. (For more information, see Chapter 5.)

Obviously, only satisfactory material should be released for production.

Stampings: While inspecting stampings, one must look out for signs of tool wear, tool breakdown, rough handling, etc. By the way, the blueprint does not always tell the whole story. The detailed specifications or instruction sheet that accompanies the production line-up of the stamping must be taken into account as well.

(*Note.* In order to avoid misunderstandings, it must be stated that this chapter covers only such defects as may occur after the start of the regular run, i.e., after the first samples have been passed by the proper authority. Excluded are defects that may be caused by poor material, poor tool design, poor tool construction, incorrect die setting, wrong stamping technique, erroneous stamping design, etc.)

First of all, dimensional accuracy must be ascertained. The dimensions and the shape of the part must correspond to those given by blueprints or gages or master samples. Tool wear causes variations in part characteristics, and these variations should be detected before they pass the allowed tolerance limits.

While, in machining, the dimensions may vary from one workpiece to the next, because of lack of rigidity of the machine tool or the tool holder or because of tool wear, in the production of stampings the trend is rather uniform and unidirectional: it is due to natural wear by abrasion of the chief tool members—punch and die plate—which results in the following: blank sizes (outer contours) increase, hole sizes decrease; and bent angles become bigger. In addition, certain tool defects, such as worn or deformed stop pins, pilot pins, and stock gages in progressive dies, cause incorrect location of holes punched in preliminary stations, in relation to the blank contour.

There are certain press tools with adjustable members that have a direct influence on the dimensions of the parts. A typical example consists of parting dies where the adjustable stop governs the length of the workpieces. The loosening or wear of such tool details may cause dimensional changes.

With progressive dies and multiple-punching dies, one must check whether *all* the holes in the stampings are perforated; if a hole is missing, the corresponding perforating punch may be broken or worn too short, and/or the shut height may have become too high through loosening of the pitman lock nut and unscrewing of the pitman screw.

Flat Blanks: Examination of the shear marks and of the break on the sides indicates a great deal to the experienced inspector. If there is one side with longer shear marks and less break then the opposite side, it means that the punch-to-die clearance is not evenly distributed (punch and die plate alignment having shifted somewhat) or that the die plate is not perfectly square with the punch. Even with correct and evenly distributed clearance, it may happen that some section of the side is "fired," i.e., scratched, scored, or burnished. This can be caused by metal particles adhering to the die plate opening walls.

If the stock is too thin to observe, then stack together a few stampings in the same order as blanked. In such a way, the examination becomes much easier.

Blanks and punched holes always have some burrs around the periphery. Burr height should not be excessive; it must be small enough to be removed by tumbling or some other economical method, if it must be removed at all. Too great a burr is a sign of dull tool cutting edges, worn tools (increased die plate opening and/or decreased punch size), or a cracked die plate.

For the determination of the admissible burr height, the function of the blank must not only be taken into consideration, but often all further processing of the blank until it becomes a finished component or product. For example, in formed or drawn parts, if the burr on the primary blank is excessive, the forming or drawing tool may easily become damaged and, consequently, have to be restored too early and too often.

Not only the size of burrs is to be inspected, but also the side on which they lie. Blanks destined for further forming operations must have the burr on the side of the least stress, usually the interior of the formed stamping.

Flatness of blanks is essential in many cases. Distorted blanks are due to twisted strips or coils or to damaged cutting edges of the die plate and/or of the punch.

Causes of die breakdowns are so varied and numerous that it is impossible to cite them all here. At any rate, the consequences of such trouble are quite evident and typical, and the operator should be able to recognize them immediately. For example, if cutting edges of the punch or of the die plate are nicked or chipped, the blanks reveal such a problem by marks and bends. In addition, as mentioned above, damaged cutting edges on the die plates usually produce correspondingly heavy burrs in the blanks, and a chipped punch usually leaves a ragged shear mark on the blank.

Progressive dies equipped with pilot pins may damage the punched pilot holes, if the feeding is not performed properly.

Small slugs sometimes return from the die opening, and are deposited either on the blanking punch bottom or on the die plate top ("slug pulling"; see Chapter 15). In subsequent blanking or punching, the slug leaves a distinct indentation on the stock.

Among stock defects, there is one which may escape the initial stock inspection and become evident only during production. This is the so-called "lamination" caused by impurities embedded in the stock. These cause splitting of the sheet metal along the sheared edges, and occurs chiefly with brass, bronze, and nickel-silver, and sometimes also with steel.

Delicate surface finishes that must be preserved are liable to damage by careless handling throughout the whole processing operation. In forming operations such problems may be caused by foreign matter and by scored tool members. Dirt, scale, and other foreign matter may be found on the stock, in the blank-carrying bins, on the tool, in the lubricant, on the operator's hand or gloves, etc. The die ring of a drawing tool may become scored by any of the just mentioned causes and also become "loaded" by some picked-up stock particles. Again, die plate openings of cutting tools may also become "loaded" causing scratched or badly burnished blank sides. Such surface imperfections must be eliminated later by operations that considerably lower the overall efficiency and economy of the production.

With second operation cutting dies, inspection must be even more careful than for first operation tools, because second operation tools entail a much higher danger of dimensional inaccuracies and/or burrs on the wrong side. Apart from the unavoidable human factors (e.g., wrong feeding of symmetrical blanks), there is the possibility of inaccuracies of the blank-locating nests caused by wear or loosening and shifting; by dirt, chips, or foreign matter; or by a change in the alignment of the blanks in the nests.

In first operation tools, it is advisable to examine not only the blanks, but also the strip skeleton. The appearance of the blank openings and that of the remaining stock surfaces are excellent supplementary clues about problems in the tool.

With flat blanks destined for certain forming and bending operations, the rolling grain direction must be checked. (It should lay across the forming or bending lines.)

Formed parts: Bent or formed stampings must be inspected more frequently than flat stampings, because forming dies more easily produce dimensionally inaccurate stampings than do cutting dies. Variations in physical and mechanical properties of the stock, in thickness, and in temper of the sheet metal shift the location of bends and the magnitudes of the bent angles (springback phenomenon). As with second-operation cutting dies, in forming dies inaccuracies of nesting and/or wrong feeding of primary blanks may cause defects. All this demands much closer dimensional control of formed stampings.

The chief observable defects that may occur in forming operations are pressure marks, cracks, and indentations.

Marks of pinching, due to excessive forming pressure (e.g., caused by too deep tool setting or by loosening of the ram-adjustment locking nut) are not only unsightly, but also often impair the strength of the part, because of highly concentrated stresses in the bending zone.

Cracks in bends and corners also mean a serious decrease in part strength. In an otherwise approved part, it may stem from the omission of a specified, necessary annealing operation or the use of improper or insufficient lubricant or the use of inadequate raw material. In a new part, cracks may mean too small a rounding radius of the corresponding male punch.

Analogously to what happens to flat blanks, some indentations may be found on the finished workpieces of formed parts. Such marks are made by deposited foreign matter or chipping of the active surfaces of the main tool members or by scored working surfaces of the male and/or female stamps.

Drawn shells: The drawing of shells is still a matter of skill. This means that the success or failure of a drawing operation is still subject to many unpredictable factors and too many possibilities for trouble. The following covers those defects most likely to happen:

Dimensional inaccuracies: shape and/or size. *Cause:* worn tool members.

Cracks, rupturing and burst shells. *Causes:* excessive blankholding pressure; too high press-ram speed; shifted punch alignment (either nonconcentric or inclined punch axis with respect to die opening axis); incorrect blank location (nest worn or loose; presence of dirt, chips, foreign matter, etc.); improper lubricant.

Wrinkles. (1) All around the free top end of the shell. *Causes:* too low blankholding pressure; too heavy or too light lubricant (not slippery enough or too slippery); worn tools. (2) Only on one side of the shell. *Causes:* Uneven blankholding pressure; incorrect location of blanks in the nest; primary blank with burrs on one side.

Uneven shell height. *Causes:* incorrect (nonconcentric) nesting of the flat blank; uneven distribution of the blankholding pressure; blanks with uneven burrs; shifted tool alignment.

Burnished shell walls on one side only, or in severe cases, fractured shell walls. *Cause:* Displaced drawing punch (misalignment).

Bulged shells. *Causes:* inadequate air vents; damaged or worn drawing edges.

Scored shells. See above under "flat blanks."

Problems Caused by Press Vibration

Occasionally, but more frequently with open dies (those not mounted in die sets), the continuous vibration of the press caused by the stamping impact, may loosen one or both of the tool members. Even if the punch and/or die plate does not break, there is almost always deterioration in product quality. In fact, the alignment between punch and die is altered, causing the following defects:

1. In cutting dies, the burr becomes irregular, i.e., excessive at one side.
2. In drawing dies, there may be tearing of shell bottoms, uneven shell height, or shells with fractured walls.

Sometimes unprovoked vibrations in the working of the blankholding equipment may cause serious trouble in drawing dies: tearing of shell bottom (excessive pressure), wrinkles (too low pressure), uneven shell height (uneven pressure distribution).

Inspection Methods and Means

Inspection plan: For unusual, complicated stampings, it is good policy to prepare proper plans of inspection. The inspector, the die setter, and the press operator must receive a detailed set of procedures (operation chart or sheet) which should show the chief dimensions and essential characteristics of the stamping and the critical points liable to give trouble, which must be checked more frequently and more carefully; how to inspect the stampings correctly; when to do it; and with what means.

Sampling: Unless 100% inspection is specifically demanded, random sampling inspection (statistical quality control) is the usual procedure. It is based on the following considerations.

If a certain number of samples is taken at random from a given batch of products, manufactured together under the same conditions, and if a certain, comparatively very high, percentage of these samples is all right, then in all probability, the remainder will be acceptable.

The number of samples varies from case to case. Logically, the better the tools, the machines, the equipment at large, and the operators, the smaller may be the sample lot. On the other hand, the nature of the part must also be taken into consideration. This means that the number of samples must be increased with higher quality requirements, i.e., with closer tolerances.

The interpretation of the sample checks is different from plant to plant and even from stamping to stamping. Usually it is said that if about 80% (some inspection departments are more strict and demand 90 or even 95%) of the samples is all right, then the whole batch is accepted; otherwise, the lot must be subjected to a second, more severe, sampling inspection or to outright 100% inspection. Finally, there are also more or less complicated inspection procedures based on mathematical sampling (see Chart 12-1).

Chart 12-1: Scientific Sampling Method

From every lot of 1000 units, take three groups (at random), consisting of three, six, and ten units, respectively.

a. Check the first group of three units. If all of them are acceptable, then the whole lot is accepted without further checking.

b. If one unit of the first group is wrong, then the second group of six units is checked. If all six of them are acceptable, then the lot of 1000 is accepted without further checking.

c. If one unit of the second group of six is wrong, then the third group of ten units is checked. If all of them are acceptable, then the whole lot of 1000 is accepted.

i. In case any one of the third group of ten should fail, the whole lot of 1000 is rejected.

ii. If from the second group of six units, more than one is wrong, the whole lot of 1000 is rejected.

iii. If from the first group of three units, two are wrong, then both the second (six units) and the third group (ten units) must be checked. If one single unit of these 16 fails, the whole lot of 1000 is rejected.

iv. If all three units of the first group are wrong, there is no more checking and the whole lot of 1000 is rejected.

Inspection aids: The inspection tools and aids must, of course, always be adequate for the task to be performed (i.e., they must be selected according to the specified accuracy grade and other characteristics of the stamping), and must be in excellent working conditions.

The actual control may be performed either by measuring or by testing. In the first case, the dimensions are determined with measuring instruments; in the second case the stampings are compared with gages or master stampings.

Generally, in machine shops and press shops, measurements are concerned chiefly with length, width, height, thickness, hole diameters, distances from formed details, spacings between hole centers, location of holes, squareness, angularity, concentricity, flatness, alignment.

According to the tolerance limits specified, the measuring activity may be performed either with common shop tools and instruments (such as scales, steel rules, squares, straight-edges, vernier calipers, micrometers, feeler gages, bevel protractors, radius gages, amplifying dial indicators) or, for higher accuracy measurements, with sophisticated equipment (such as microscopes, optical comparators, coordinate measuring machines).

It is sound practice for the measuring equipment to be accurate to the next decimal place beyond that specified on the part print.

For testing, such standard tools and devices are usually employed as one-dimension plug gages, ring and snap gages, or two-dimension limit gages of the same design, i.e., tools with fixed magnitudes. For closer tolerances more sophisticated devices are used, such as dial indicator limit gages, amplifying comparators, optical projectors, coordinate measuring machines, as well as contour gages for checking formed parts.

In addition, in order to get the quickest and best results, specially designed appliances and devices are employed, often consisting of quite simple tools. For example, in the case of punched panels, where a rather large number of holes are punched simultaneously, it is very difficult to check visually whether or not all the holes have been punched. For checking such panels, it is common practice to prepare a wooden board with small nails driven into it at the proper locations. Putting the stamped panel over the nails demonstrates instantly whether a hole has been punched (Fig. 12-1).

Functional gages: The consequences of mechanical short circuits or breakdowns are easy to detect without special inspection equipment. However, such devices are needed to inspect those details which may imperceptibly change the dimensional accuracy, because of natural wear of press tools. For example, to test the distance of holes to certain outer contour details, or a simple case of go/no go, a gage is made with a round shaft (diameter somewhat smaller than the hole diameter) which carries a flat steel plate correctly shaped, with the corresponding *stepped* maximum and minimum dimensions (Fig. 12-2). Functional gages are better: here the stamping is located on a suitable stationary pin of correct size (simulating the actual pin on which

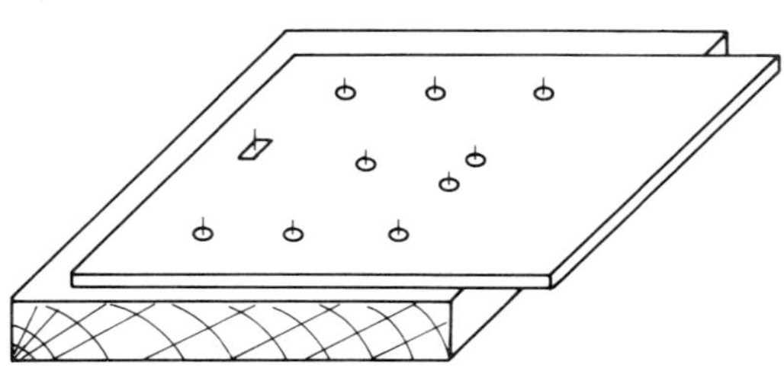

Fig. 12-1. Simple fixture for checking the presence of a large number of holes punched in a panel.

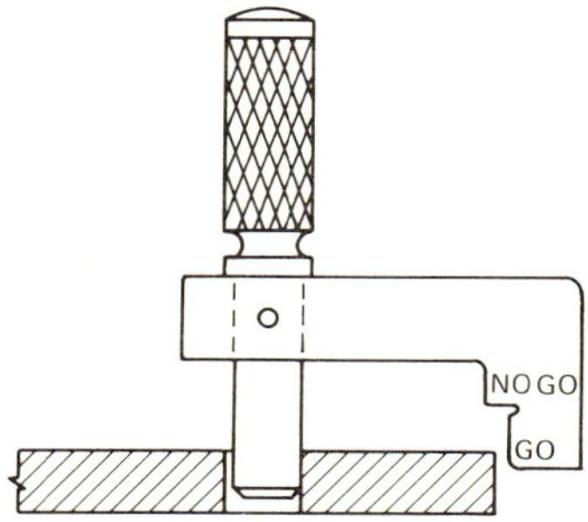

Fig. 12-2. Functional go–no go gage for checking the distance of a hole from an edge.

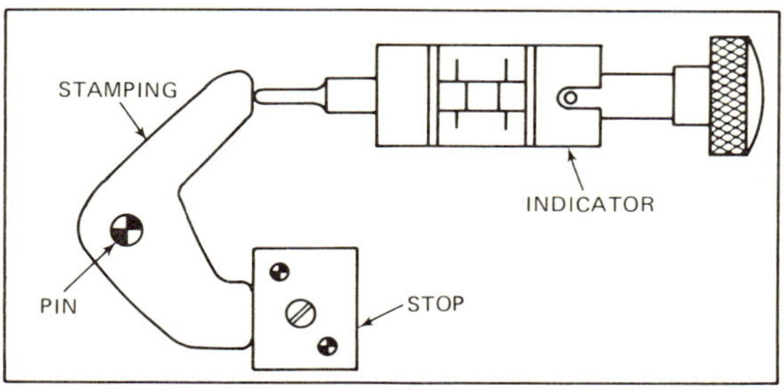

Fig. 12-3. Direct acting homemade functional gage for large tolerances.

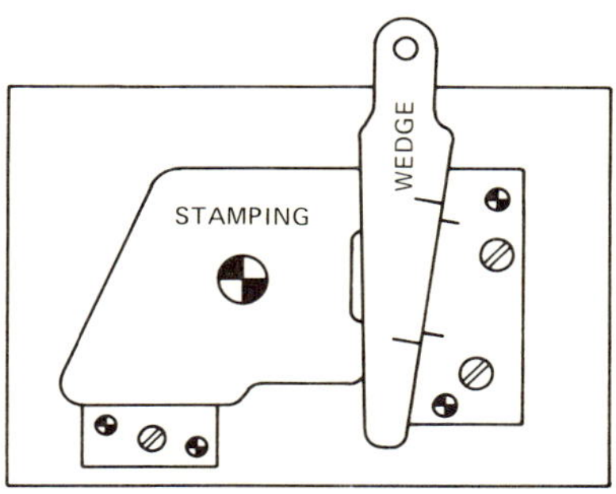

Fig. 12-4. Functional gage with tapered key for closer tolerance checking.

the stamping will be located in the finished assembly) and at the right distance a tester controls the stamping details. For large tolerances (± 0.010 inch or more) simple, homemade go/no go direct indicators may be used (Fig. 12-3). For closer tolerance limits, a tapered key may be utilized, which amplifies the indications of the tester (Fig. 12-4). For even higher accuracy (smaller tolerances), amplifying dial indicators must be employed (Fig. 12-5).

The distances of holes from each other and the blank contour are tested with a gage similar to that of Fig. 12-6. The testing pins are the size of the minimum tolerance limit while the stops for the outer contour are located at the maximum tolerance limit of the blank.

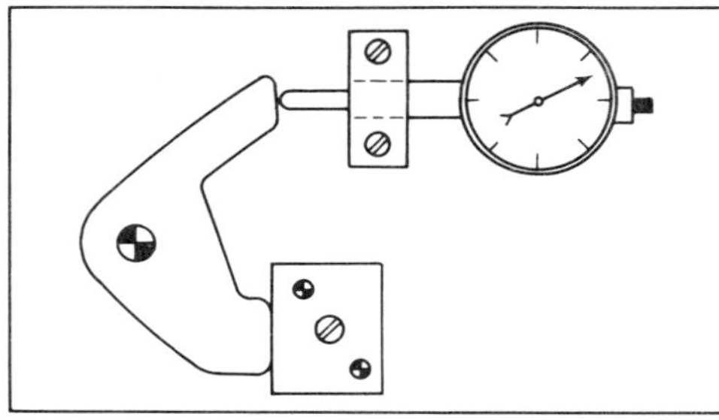

Fig. 12-5. For finer tolerances dial indicators are used in functional gages.

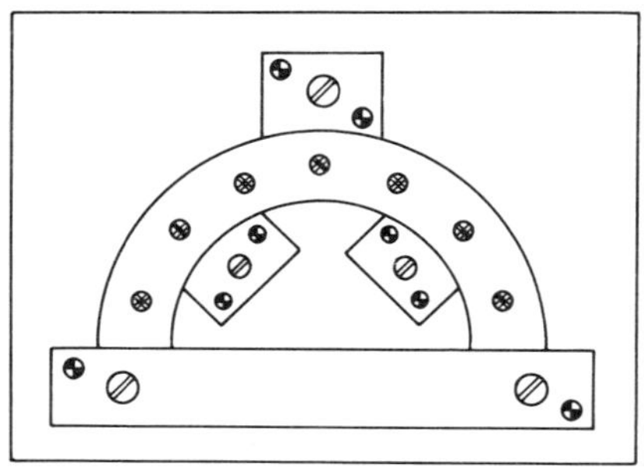

Fig. 12-6. Checking for hole distances to each other and blank contour.

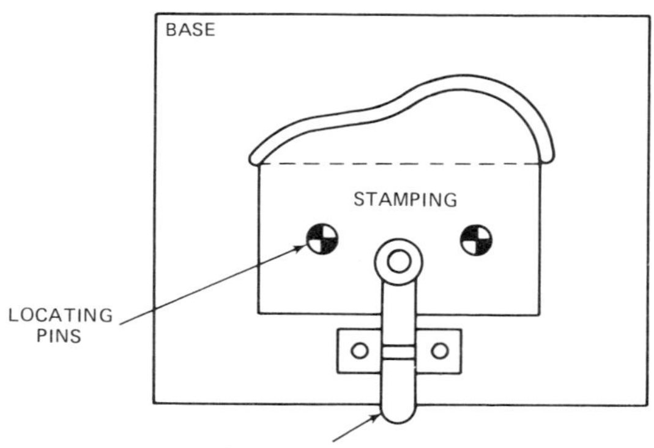

Fig. 12-7. Checking fixture for specially formed stampings.

For the inspection of outer contours or critical contour details and/or their locations from holes or other details, profile gages, templets, and optical comparators are used extensively. Figure 12-7 shows a contour-checking fixture that works in combination with go/no go plug gages (not shown) which checks the gap between the stamping contour and the inner contour of the fixture.

The determination of burr height is a rather delicate procedure. Usually, it is left to the judgment of the inspector but this is not good practice. The burr height should

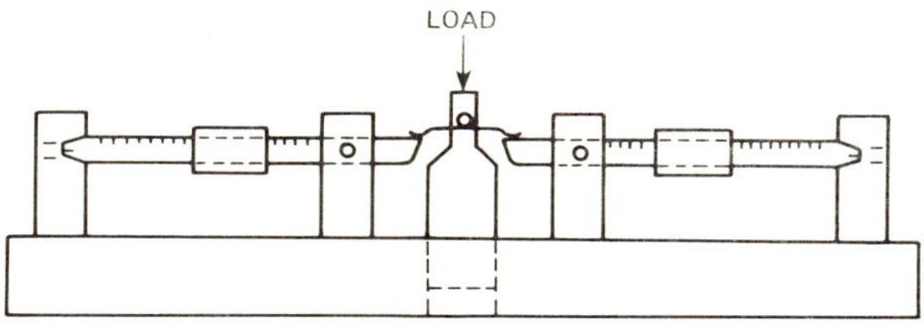

Fig. 12-8. Functional fixture for checking formed flat springs.

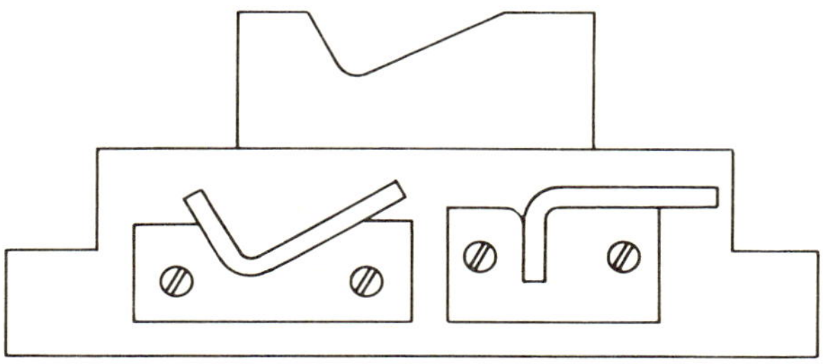

Fig. 12-9. Gages attached permanently to die shoe for checking formed stampings.

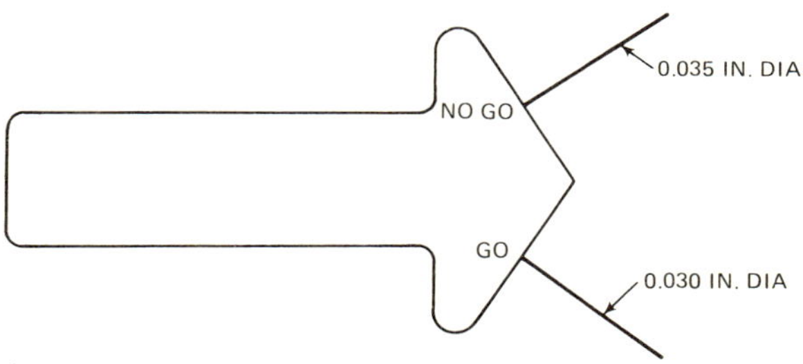

Fig. 12-10. Go–no go checking of small gaps with gage pins located in common holder.

actually be measured but not with ordinary micrometers; these flatten the burr and give incorrect values. There are devices on the market that measure the burr height without deforming it (guesswork and personal judgment are eliminated). Such commercial devices usually work on optical or mechanical principles.

Springs formed from flat material, such as are used widely in electronic devices, must be inspected not only for dimensional accuracy, but also for correct performance. In fact, the slightest difference in analysis or temper of the raw material produces considerable changes in the behavior of the springs. A very simple, homemade inspecting device for formed flat (double-sided) springs is shown in Fig. 12-8. The spring is put into the device in the same position as it would be in the final assembly. When loaded with the pressure it will be subjected to in its job, two arrows—at the end of the scale-arms—indicate whether or not the spring tension is between the required tolerance limits.

Sometimes the inspection is combined with some adjusting operation, especially in case of high-quality, small, delicate, formed stampings. These are inspected in

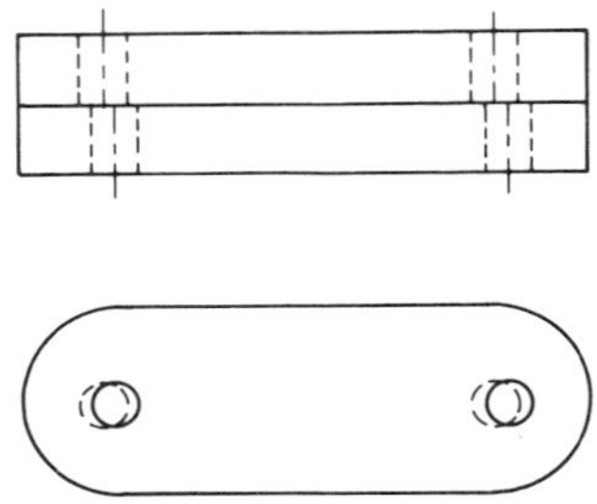

Fig. 12-11. A "trick" for easy checking of punched hole location in stampings.

special devices in which they may also be adjusted, if and when their characteristics do not conform to specifications.

Practical Hints

Gages for the control of bent or formed stampings are best fastened to or held on the stationary member of the die. In such a way, the gages are not easily lost and are always at hand for the operator who must control the exactness of the workpieces during the production run (Fig. 12-9).

In order to expediate 100% checking of a small critical gap in a stamping, the following simple device was made (Fig. 12-10): a holder was prepared from ¼-inch-thick aluminum plate. Standard calibrated, hardened, and ground gage pins of the right diameters were put into previously drilled holes on the inclined surfaces.

To check symmetrical stampings produced with progressive dies, cement together (or at least hold firmly together) two reversed stampings so that their outer contours coincide perfectly. Any error in the location of the punched holes is thus doubled and becomes apparent immediately (Fig. 12-11).

In case of job stamping plants, it is good policy to ensure that the inspection methods and gages used in the production plant are identical with those the customer will use when performing the receiving inspection.

It is well to keep in mind that the common practice of purchasers of stampings consists of accepting any reasonable part (even with slight deviations from print and/or specifications) when they are in a hurry, but of doing detailed inspections when their inventories are high. In all relationships between the customer and the supplier of metal stampings, the blueprints are only as binding as the customer wants them to be!

CHAPTER THIRTEEN

Accident Prevention in the Pressroom

Press operators are among the industrial workers most exposed to the danger of accidents, owing to the hazardous nature of the work they perform. According to reliable sources, the percentage of accidents that happen to people working with presses—accidents which result in permanent disability of 20% or more of the time—is higher than the *total* percentage of accidents with all the other standard machine tools (drill presses, lathes, milling machines, grinders, etc.).

The problem of safety, or more correctly, the prevention of accidents to the operators and other employees working in the pressroom, is of utmost importance. This problem—of course, not only in pressrooms, but also in any kind of industry and in any kind of activity—has a triple aspect: moral, social, and economic.

In the first place, it is the responsibility of management to prevent every possible injury and permanent disability to their employees while on the job. In the second place, an injured worker cannot work for a more or less long period of time, often causing serious emotional and economic problems for himself and his family. Finally, there is also the economic factor for the company. Aside from the direct financial losses consisting of compensation (indemnity payments or increases in insurance rates) there are also hidden costs, such as: (1) Decrease of production output because of the forced layoff of a trained worker (and training of a replacement). (2) Loss of actual production by other workers caused by fright, empathy, or curiosity and disturbances provoked by the accident. (3) Loss of time by other employees, foremen, supervisors, etc., while assisting the injured person, investigating, and reporting the accident. (4) The bad feeling—lasting from several hours to many days—of the employees who attended the injured person (in small shops it is the foreman, supervisor, or simply a fellow worker who gives first aid). (5) Costs of repairing damaged tools and damages to work and to equipment. (6) Lowered efficiency of the injured worker when he returns to the shop. (7) Loss of customers because of delays. (8) Generally lowered morale among the workers. All of this means a productivity decrease.

It is es)imated that such hidden costs amount to three to five times the direct, actual costs of an accident.

Satisfactory safety in a given shop *does not happen spontaneously*. On the contrary, everyone connect with the job must *work hard* to achieve it. Only through care and the good will of all concerned may both the number and seriousness of accidents be avoided completely or at least reduced to a minimum. It is a special cause for satisfaction that during the author's professional career there were no serious accidents in the press shops managed by him. This record was achieved through following a policy which is the result of extensive professional experience in this field.

The whole problem of industrial safety consists chiefly of prevention or anticipa-

tion. Management must provide methods and means which will effectively protect the workers. In the specific case of press operators, such methods and means must be concentrated especially on protecting the fingers, hands, and arms of the workers and must prevent any possibility of injury to them. Of course, it is imperative to prevent any chance of other accidents provoked either by mechanical or by human causes.

Human Causes of Accidents

It seems incredible how important a role is played by the human element in causing accidents. According to statistics, about 85% of all industrial accidents are caused by unsafe acts (human factors) and only about 15% are caused by unsafe conditions (mechanical failures).

Actually, this percentage ratio would be even worse, if we took into consideration the fact that behind the majority of unsafe conditions is the human factor responsible for the lack of proper care and maintenance of equipment.

The list given below is taken from reports of accidents that actually happen. Almost everyone connected with press working is able to confirm many of these causes from his own experience and, perhaps, add some more. The human causes may be classified, according to their nature into a few groups:

1. Physical causes: Tiredness, fatigue, lack of sleep, illness. Fatigue is not always simple physical tiredness. More often it is combined with mental and intellectual tiredness. This is due to the extremely repetitive character of press operations. In fact, it is obvious—and actual experience confirms it—that the human mind is not capable of performing 15,000–20,000 identical operations in a day without one distraction. And just one such careless instant may cause a finger or a hand. Also a high noise level may cause fatigue.

2. Involuntary actions. A typical case is an accident that occurs when a workpiece is put in the nest of a die in a wrong position or the workpiece comes out of the nest before the operation is finished. In such cases, it is instinctive to reach for the part to rearrange it correctly. And then an accident happens.

3. Another analogous cause is related to the fact that the attention of the press operator must be concentrated on his job. If we compel him to be doubly careful, i.e., attend properly to his job and at the same time also to his safety, we are asking too much.

4. Causes depending on the operator's state of mind in a given moment. Worries (domestic, health, financial, or other preoccupations); grudges against some other people; misunderstandings with the boss; low morale; etc.

5. Lack of training and/or lack of discipline: inexperience or ignorance; an attempt to relieve a jam; to remove scrap with hands, instead of using a stick or a pick; trying to clean, lubricate, or adjust the press while it is in motion; trying to prepare the press for the next task (while the machine is in motion); trying to "beat" the rhythm of the press; lack of attention, alertness, or thoughtfulness; distractions or lack of concentration; lack of interest in the job; lack of good will; poor housekeeping; curiosity; overconfidence; showing off; recklessness; imprudence; and horseplay.

6. Semisabotage. In order to reduce hand movements or for increasing produc-

tion rate (operator on piecework or on incentive). Deliberate disregard of specific safety rules and instructions and putting safety devices out of order.

Mechanical Causes of Accidents

These may be classified roughly, according to their nature, into two groups: avoidable and unavoidable.

a. Avoidable causes are those that should not happen; they can—and must—be prevented: lack of or ineffective safety guards, press in unsatisfactory condition, "frozen" pulley on the shaft, broken clutch dog, broken or loose clutch springs, loose clutch and/or brake, defective electrical installation causing shock.

b. Unavoidable causes cannot be prevented, they just happen: flying pieces of stock and/or of tool members, which, when broken, injure the face, eyes, or neck of the operator; breakage of the crankshaft and thus dropping of the press ram; breakage of the shaft with flywheel falling down.

However, as mentioned before, a well-organized preventive maintenance program, together with alert die setters and press-shop foremen, can prevent many of these so-called unavoidable causes, or at least the severity of their consequences can be reduced considerably.

Consequences of Accidents

In hand- and foot-powered presses, when the operator becomes aware of the beginning of an accident, he can often stop the application of pressure, thus avoiding a severe injury. Therefore, for these kinds of presses, safety measure are usually not provided. However, in the author's opinion, this is a great mistake, because accidents in hand and foot presses often do happen and not too rarely (in about as much as 20% of the cases) serious ones, with permanent disability as a result. One must take into consideration the fact that in big flywheel presses, the kinetic energy prevents stopping the ram movement quickly.

Mechanical presses are, of course, more dangerous. Because of their high speed, if a mistake is committed, it is impossible to intervene in time to reduce the consequences of an accident. They are almost always serious. In addition, it should be remembered also that a small mechanical press flattens fingers with the same ease as large and powerful ones, and the consequences are just as bad in both cases.

Basic Concepts of Pressroom Safety

As mentioned above, the goal of accident prevention consists of avoiding the possibility of injuries to press operators. Such a program produces an atmosphere of safety that increases confidence and thus also production rates.

Pressworking of metals is an ancient art, several centuries old. However, people have become safety conscious only in the last few decades. (Up to the beginning of the century it was simply taken for granted that accidents with presses happen. A press operator with "only" one finger missing was considered to be rather skilled.) Finally, as usually happens with empirical solutions, people began to combat the symptoms and not the causes of the problem. First appeared the protective safety appliances for preventing accidents, and these still enjoy a certain degree of popularity. However, although they do some good in certain cases and under given circumstances, these devices do not represent the *true* solution for the problem of accident prevention in the pressroom.

Recently, safety experts all over the world began to realize that true, effective accident prevention, which had a very high proability of success, can indeed, be achieved. But it is done through specific technical and scientific procedures and by applying the following, simple reasoning:

First, any *need* for the operator to reach with his hand or fingers into the danger zones ("under-the-punch"; "point of operation"; in a "nib point," "pinch point," "shear point," etc.) either for placing or removing parts or anything else while the flywheel of the press is in motion must be strictly avoided. This rule is also valid for die setting and any other operational adjustment, repair, maintenance, or servicing.

Second, at the same time the remotest probability for an involuntary accident to the operator, i.e., the *possibility* of his fingers entering into the danger zones, must also be avoided.

These two conditions are fulfilled by the following means:

a. Adequate tool design plus proper—indirect, semiautomatic, or fully automatic—feeding and ejecting devices solve the first condition.

b. Point of operation guards on dies or on presses take care of the second condition.

c. Additional help is given by auxiliary equipment and by proper training and indoctrination of press operators, die setters, and other workers.

By the way, the following axiom should be kept well in mind: *The three concepts of Tool-design, Safety, and Productivity are intimately connected.* Any one cannot be ignored without seriously affecting the others.

Safe Press Tools

As was mentioned in previous chapters, press tools are classified according to the kind of operation they perform and to the method of production. However, from the safety standpoint, the classification divides the dies simply into first and second operation tools. The safety approach is fundamentally different in each case. It is the purpose of the following to show how a standard press tool can be rendered "safe."

The easiest types of tools to make safe are first operation dies. In such tools, the first safety condition is automatically and intrinsically fulfilled: there is really no need for the operator to reach with his hands into the interior of the tool while performing his job. But, as already stated above, it is not sufficient to avoid the *need* for putting the hands into the danger zones; it is also necessary to avoid the *possibility* of doing so. This second safety condition is fulfilled by rendering the tool thoroughly safe through the following method, which prevents accidents even in many cases of "unavoidable" mechanical failure or breakdown such as, for example, shaft breakage:

1. Giving preference to certain types of die construction,
2. Modifying more or less substantially standard tool design details, and/or
3. Adding proper guards around the tool.

Danger Zones in Dies

Speaking in terms of die anatomy, the actual danger zones in a first operation die must be determined. (By the way, in the other die types, there are practically the same danger zones.)

There are several danger zones (refer to Fig. 13-1): (1) between punch(es) and die plate; (2) between punch(es) and stripper; (3) between punch holder and stripper; (4) between parts of the die set; (5) between accessories.

In dies with solid, stationary stripper plates (preference should be given to this type of stripper design, because "closed" dies are much safer *per se* than dies with movable strippers), accidents may happen at both sides of the stripper plate: at the top and at the bottom, danger zones (2) and (1). The latter is easily taken care of (Fig. 13-2): The height of the channel where the strip moves, between stripper plate, die plate, and stock gages, should be less than ¼ inch, so that there is no possibility of introducing the fingers into it. (By the way, for cutting tools stamping thin or medium gage stock, a ¼-inch channel height would usually be excessive, anyway.)

In order to prevent the possibility of reaching between the bottom of punch and top of stripper, danger zone (2), proceed in this way: Use the shortest practical ram stroke and thus prevent the punch from leaving the stripper. To permit this, the stripper plate should have sufficient thickness, i.e., somewhat more than the usual values. If it is not practical to do so (i.e., to increase the stripper plate thickness), then provide some suitable auxiliary tubing (Fig. 13-2). In those cases where protecting prolongations are not practical (e.g., long strokes are needed for performing the operation) and it is unavoidable that the punch leave the stripper, proper guards around the punch must be provided. The best guard designs are: (a) helicoidal or telescoping springs for long or short ram strokes, respectively (Fig. 13-3), with the opening between two adjacent turns should not be larger than ¼ inch at the open die position; (b) telescoping tubings made from short pieces of pipe with flared-in and flared-out ends (Fig. 13-4); (c) leather or plastic bellows or rubber tubing; (d) stationary combination guards (see below).

To eliminate finger crushing hazards between punch holder and stationary stripper, danger zone (3), there are a couple of details the tool engineer may apply. If possible, the punches should be so long that at closed tool position there is a minimum distance of 1 inch between punch-holder bottom and stripper top (Fig. 13-5, side A). Take into account the need for future resharpenings, so start with an original length that includes an adequate sharpening margin or allowance. By the way, if the stationary part of the die is fastened to the bolster with straps that hold the die by its top surface, then the minimum distance of 1 inch should be measured from the strap surface (see Fig. 13-5, side B). Sometimes, in order to avoid punches that are too long,

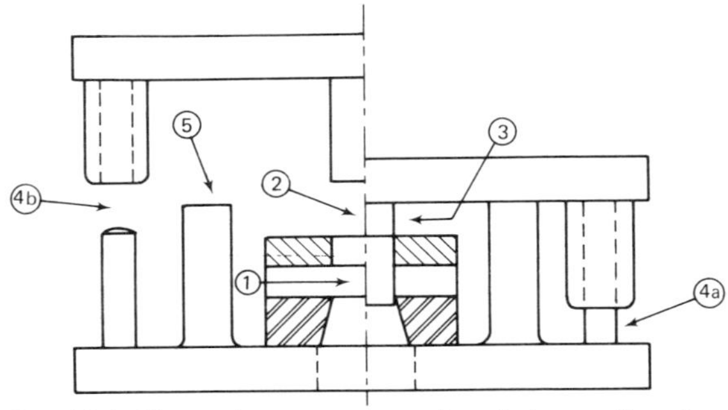

Fig. 13-1. The main danger zones in a first operation die.

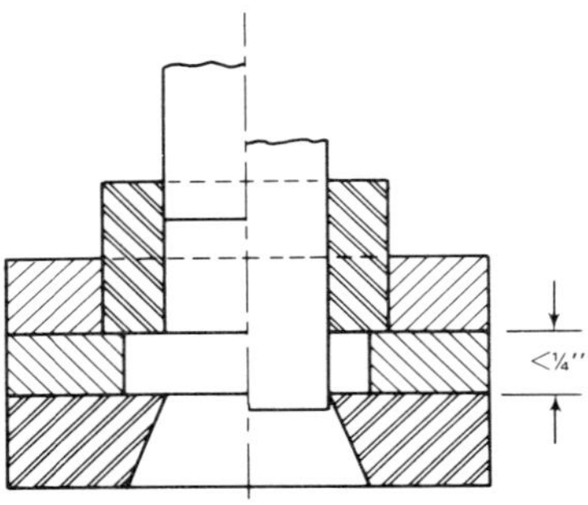

Fig. 13-2. With dies having stationary strippers make the stock gage height less than ¼ inch.

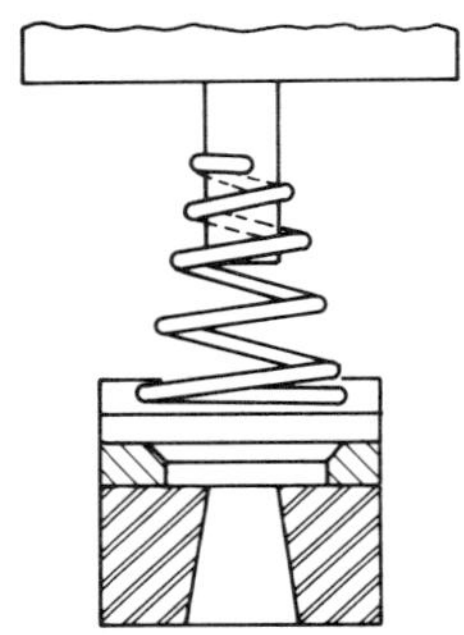

Fig. 13-3. Punch guard made from helicoidal spring.

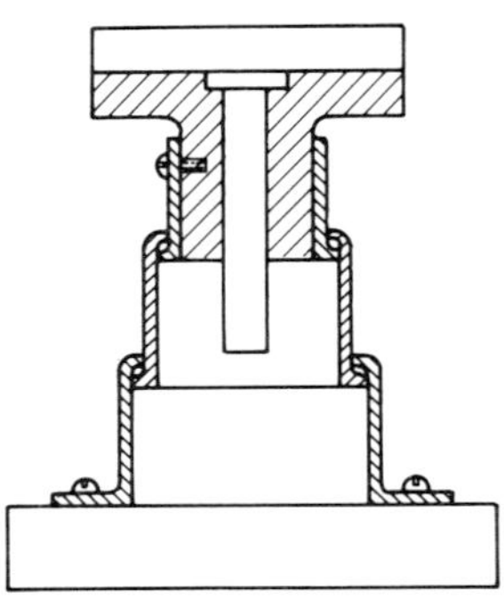

Fig. 13-4. Telescoping tubing protects against accidents between punch and stripper.

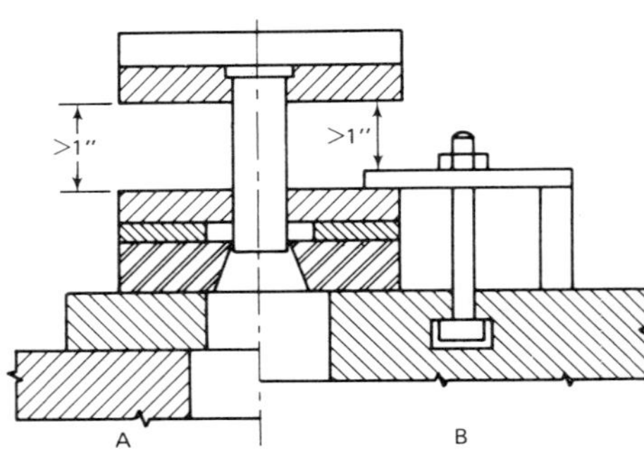

Fig. 13-5. Punch must be long enough so that there is a distance of at least 1 inch between punch plate and stripper and/or between punch plate and clamps.

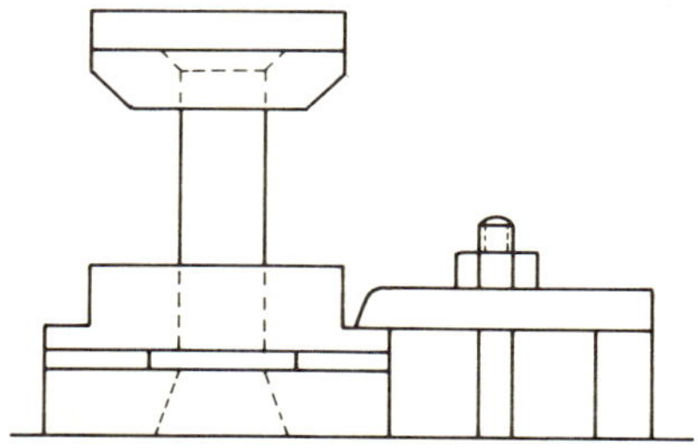

Fig. 13-6. Clamps may engage recesses in the stripper plate.

it is more convenient to make special recesses in the die plate or stripper plate to be able to locate the fastening straps at lower levels (Fig. 13-6). Here, too, is shown another, simple detail consisting of chamfering the bottom edge of the punch-holder plate, thus decreasing substantially the danger of finger crushing.

Combination Guards (Point of Operation Guards)

In many cases the safety device for danger zone (3) consists of a protective fixed guard, which surrounds the area of the punches between the punch holder and the stripper (Fig. 13-7). More often, this guard is combined with that for preventing accidents in zone (2) and becomes a *combination (point of operation) guard*. This acts simultaneously for both purposes.

The combination guard may be individual (single purpose) attached to one specific die (called a "die enclosure guard") to which it belongs, or it may be universal (multipurpose), i.e., attached to a given press (called a "fixed barrier guard" or "adjustable barrier guard"). In the latter case, the guard offers protection for any die that is mounted on the press. Adjustable, truly universal guards are adapatable between more or less ample limits, both horizontally and vertically; some types have parts that may be tilted.

The general consensus among press safety experts is that, without doubt, the most effective protection against accidents is provided under all circumstances by the individual, integral "die enclosure guard" permanently attached to the die (specifically to the die shoe or to the stripper plate). These guards, which are real armored shields around the danger zone, protect the operation under *any* conditions,

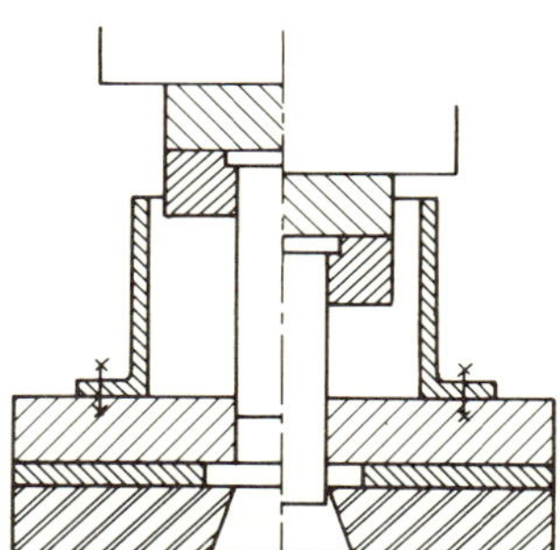

Fig. 13-7. Simple stationary combination guard attached to the stripper.

even in case of press failure (stroke repeating, ram falling down, etc.). Its advantages over the universal-type ("barrier") guard are several: It is simple in design, construction, maintenance, and is reliable in application, performance, and operation. When designed to fit the particular characteristics and requirements of the die, no adjustments are needed and are therefore not possible; consequently, the operator or the die setter cannot make it ineffective (unsafe) by erroneous adjustments. There is no possibility of operating the tool without a guard. There is no problem of storage nor of complicated recordkeeping and organizational bookkeeping. The set-up time of the tool is reduced. The operator can reach closer to the tool with safety; he can relax and he feels better.

On the other hand, against the impressive list of advantages, as outlined above, individual guards also have certain drawbacks (which the adjustable, universal-type guards do not): Individual guards are more expensive in both first cost and maintenance. In fact, the quantity of dies in any shop is always a high multiple of the number of existing, active presses; so a comparatively small quantity of universal-type guards is equivalent to a large number of individual guards. When a die is resharpened or repaired substantially, it often happens that the shut height is altered. Then the idividual guard must be laboriously modified, there being no possibility for adjustments. If the tool is used in a press with longer stroke than the maximum for which the guard has been designed, a very dangerous pinch point is created between the guard and the tool. In the case of short run or infrequent jobs, providing each die with an individual guard would be economically impractical. When developing a new accident prevention program, it would take too much time to provide each die with a guard of its own. When working with tools that belong to third parties (customers or vendors), it is not convenient, economically, to provide them with individual guards.

Design and Construction Details of Point of Operation Guards

There are several types of combination guards, individual or universal, according to particular requirements, conditions, facilities, and tradition. They may, should, or must have the following characteristics:

a. Homemade or commercially purchased. Individual guards are almost always homemade, universal guards are usually bought from outside sources.

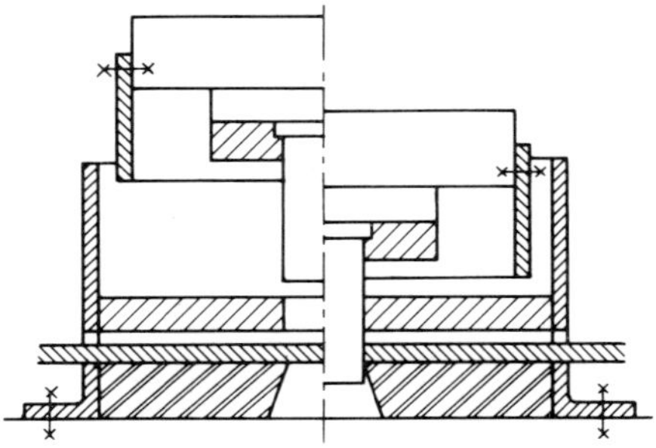

Fig. 13-8. Two-piece telescoping combination guard attached to the die plate and to the ram.

b. One-piece or composite design. One-piece guards are preferred for comparatively short ram strokes; they are stationary and attached to the die shoe or to the stripper (see Fig. 13-7). If the ram has overhanging shank-holder bolts, the guard should have a corresponding baffle plate to avoid creating a nib point between the guard and bolts. Two-piece ("wing") guards are telescopic; one shield stationary on the fixed part of the tool and the other shield attached to the punch holder or to the ram (Fig. 13-8). They are provided for longer ram strokes. (*Important note:* Stamp on the die the maximum stroke value!)

c. Transparent or opaque. Of course, the former design is usually preferred because it offers improved visibility of the work being processed.

d. Transparent guards may be made as panels from transparent solid sheets of some artificial material (nonflammable plastics, Fibestos, Plexiglas, shatter-proof nonsplitting safety glass, etc.) or as metal screens or grids.

e. "Transparent" metallic guards may be made in several ways:

1. Ordinary strip skeleton (real scrap) or custom made perforated sheet metal of comparatively heavy gage, securely fastened to a proper metal frame. The perforations are usually round or rectangular (or "obround," i.e., rectangular but with full radius rounded ends); openings should be no wider than ¼ inch; rectangular slots should preferably be arranged vertically (for decreased eyestrain and improved visibility). *Note:* Wire-woven mesh is not considered as adequate material for such guards, especially if the mesh is large. (However, sometimes such guards are used.)

2. Staples or steel rods; straight or formed (hairpin-shaped) preferably arranged vertically. These are widely used in commercially available, adjustable, universal guards.

3. Telescoping rods and tubing design, where during the whole ram stroke, access to the interior of the die is completely prevented. The best arrangement is with stationary full rods and movable tubes (Fig. 13-9). With this arrangement dirt, dust, foreign matter, etc. cannot enter the tubes and cause scoring, galling, etc.

f. A combination design is made with a plastic front panel and lateral and rear "transparent" or opaque metal shields.

g. Interlocked barrier guards are not entirely fixed; they have a side or the front movable, i.e., it can pivot or slide, in order to permit sufficient access to the die while

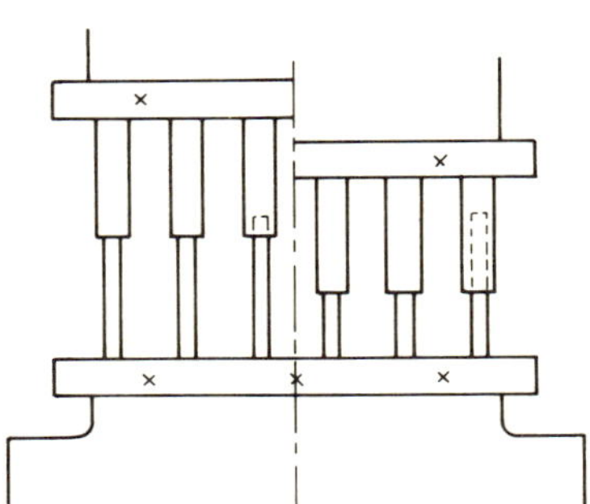

Fig. 13-9. Combination guard made with telescoping rods and tubes.

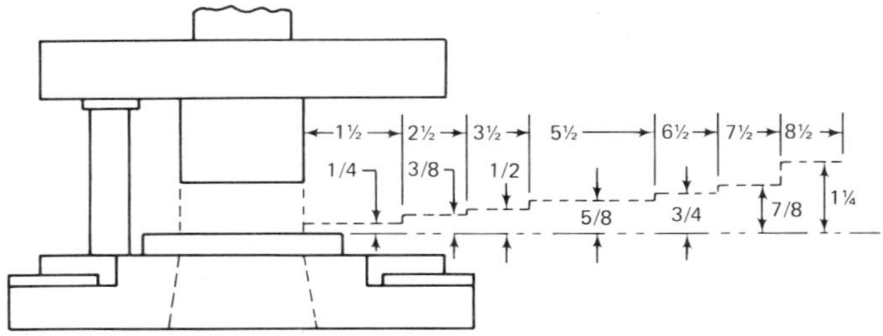

Fig. 13-10. Maximum safe width openings according to distances from nearest nip point. (All dimensions in inches.)

the press ram is in its top rest position. The movable part of the guard must be interlocked (mechanically or electrically or in some other way) with the ram control such that it cannot be put in motion while the movable guard section is in the open position.

h. All kinds and types of combination guards must be provided with properly located and adequately dimensioned slits or horizontal slots for the passage of the strips or coils (or for any feeding or ejecting device). The maximum width of such openings depends upon their distance from the nearest nip point. Conservative, practical, accepted values are given in Fig. 13-10.

i. The guard shall not create danger zones of its own, i.e., no exposed shear or pinch points are allowed between the guard itself and any moving part of the press, the die, the die set, and/or of the guard, in case of composite guard construction. By the same token, it is not permitted to use guards that, attached to the ram, may catch the operator's fingers between the gate and any stationary part of the press table (or bolster plate, die, clamping straps, bolts, etc.) in their downward movement.

In this respect, the guard of Fig. 13-11 I is incorrect, since its height is insufficient both in the open position and in the closed position of the die (A and B). On the other hand, Fig. 13-11 II shows a guard of sufficient height. Dimension z is 3/16 inch in the open position of the die.

j. The guard should not have any flexible parts.

k. It is recommended that all the outer edges of guards be rounded or reinforced, in order to avoid hand and finger injuries caused by the guard itself.

l. The coverage of the guard must be *complete,* i.e., really making it impossible

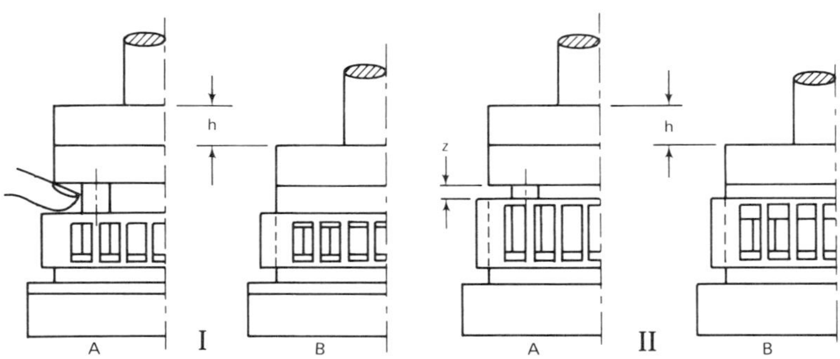

Fig. 13-11. Danger zones created by guards and how to avoid them.

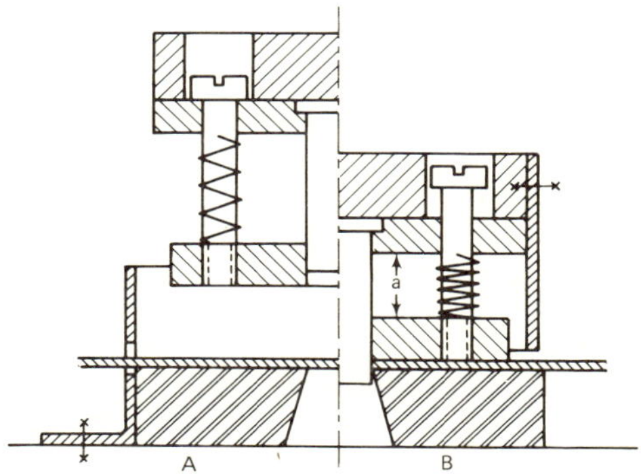

Fig. 13-12. Guards for dies with movable strippers may be attached either to the die plate or to the punch holder.

for the operator to reach with his hands or fingers into the point of operation, neither over nor under nor around nor through. The guard must cover not only the front and the sides, but also the rear and the top, of the die, if necessary (full-enclosure guards).

m. Such "tamperproof" fasteners should be employed for the construction of the guard, as are not easily removed by unauthorized personnel, so that the protection's effectiveness cannot be impaired easily.

n. The guard should be designed and built so that it is easy to inspect, check, and service.

o. In designing the guard, consideration should also be given to the economical aspect of production, i.e., it should not lower the production efficiency by interfering with the operator. Therefore, it should permit free and easy feeding of stock and removal of parts and scrap, and should allow good visibility at any moment. The nearer the guard is located to the die and to the ram, the less it disturbs the operator.

Note: There is a very important detail that must be pointed out emphatically in regard to economics: Whenever a conflict arises between high-speed production requirements and safety, without question safety must come first.

Movable Strippers

In the case of movable strippers attached to the punch holder, the following details should be taken into consideration from the safety standpoint: For preventing trouble in danger zone (1) the distance between stripper-plate bottom and die-plate top should be less than ¼ inch when the die is open. If this condition can not be fulfilled—the ram stroke is too long—then a proper stationary guard must be provided around the die plate (Fig. 13-12A). Danger zone (2) does not exist. Danger zone (3) usually does not exist; however, if distance *a* should become less than 1 inch at closed die, a proper guard must be provided (Fig. 13-12B). Of course, in case of movable strippers, combination guards are extensively used; for example, by simply combining the designs in Figs. 13-12A and 13-12B, a standard telescoping combination guard is created.

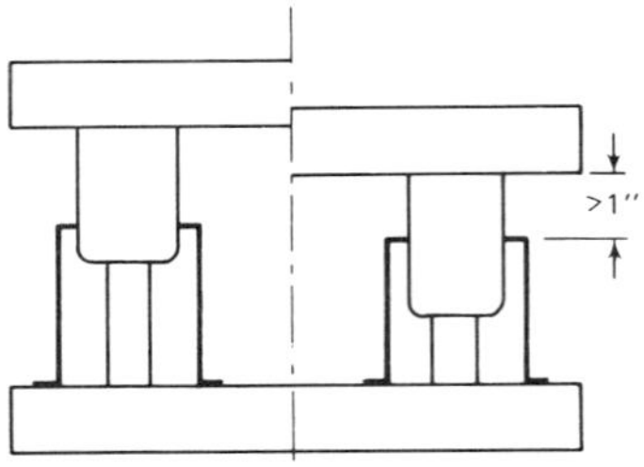

Fig. 13-13. Die-set guards must be provided if the leader pins move out of the guide bushings at the top of the press stroke.

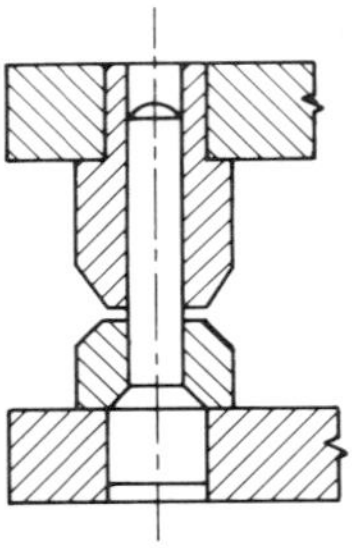

Fig. 13-14. If shut height is limited by rings, put them on the leader pins, chamfer the rings and also the guide bushing bottom.

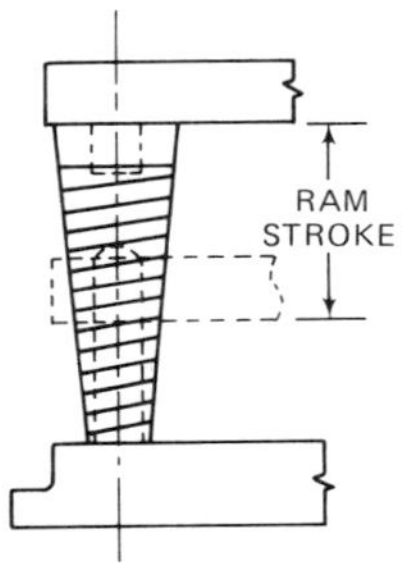

Fig. 13-15. Die-set leader pins are protected sometimes with spring leaf guards.

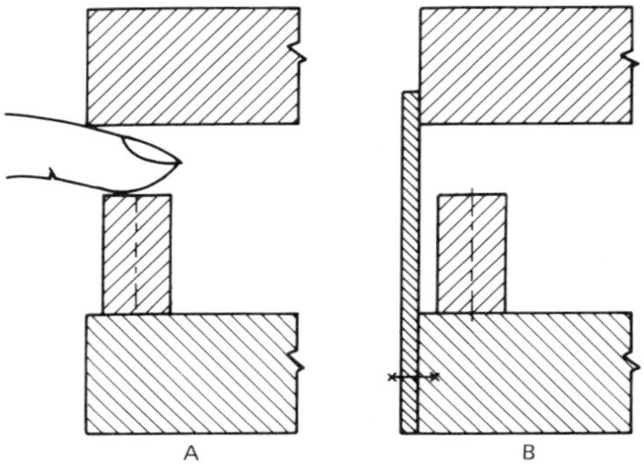

Fig. 13-16. Rectangular guard for protection against accidents with bumper blocks.

Die Sets

These often create danger zones of their own (see Fig. 13-1). There are two places where fingers may be smashed: (a) between the bottom of the bushing and the die holder (if the shut height is too small) and (b) between the bushing bottom and the leader pin top (if the shut height is too high and the ram stroke is so adjusted that the punch holder is allowed to leave the leader pins; e.g. a progressive, die with incorporated forming operations). In addition, both problems may sometimes occur simultaneously in the same die. The remedies are analogous for both kinds of problem.

1. If the ram stroke is short, e.g., straight blanking, and the dimensions of the die set, bushings, press tool are convenient, a fixed guard in the form of a tube of adequate size (loose fit to the bushing outside diameter) is fastened to the die holder (Fig. 13-13). Another solution consists of chamfering the bushing bottom and putting a properly chamfered ring at the bottom of the leader pin (Fig. 13-14).

2. If the ram stroke is too long, a telescoping guard, made from pieces of pipe of different diameters (see Fig. 13-4), or a spring leaf guard (Fig. 13-15) is used. The first is usually homemade; the second is a commercial product.

Bumper Blocks

Descent limiters and other protuberances, danger zone (5), must be adequately guarded; and, for better visibility, it is recommended that the bumper block should always be put on the stationary part of the tool or on the die holder of the die set. For this purpose, similar devices are provided as for die-set pillars: (a) tubes (like Fig. 13-13); (b) rectangular guards, just like any other similar guard (e.g., Fig. 13-16); (c) telescoping guards (like Fig. 13-4).

The devices mentioned above for accident prevention in zones (4) and (5) are used, of course, not only for first operation cutting tools, but also for any other kind of second operation press tool.

Starting Stops

These must have adequate extensions in such a way that they could be handled outside of the shield of the guard (remote control).

Springs

If in a die there are compression springs that are exposed, i.e. accessible to the operator, it is absolutely necessary to cover them with suitable closed-bottom tubes that keep the operator from any physical (accidental or voluntary) contact with the springs.

"Naturally Safe" Tools

Sometimes a die may be made inherently safe rather easily, i.e., without the need of any supplementary help in the form of guards. In Fig. 13-17 is shown such an example—taken from actual practice—where all the possible danger zones have been taken care of by the tool designer.

Zone (1) is eliminated by a channel height of less than ¼ inch (as mentioned above, this value is satisfactory for straight blanking of comparatively light gage stock). Zone (2) is avoided by making the stripper sufficiently high and the ram

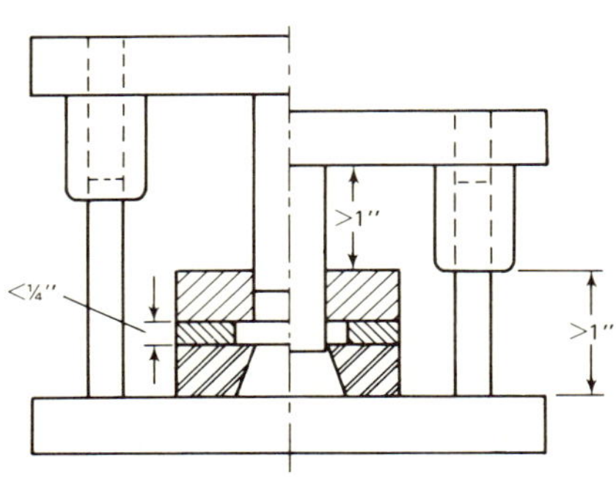

Fig. 13-17. Naturally safe tool, which does not require special precautions.

stroke sufficiently short so that the punch does not leave the stationary stripper plate. Zone (3) is eliminated by making the punch sufficiently long. Zone (4a) is eliminated by a high shut height. Zone (4b) is avoided by the short ram stroke so that the bushing does not leave the leader pin. Zone (5) does not exist, because in this case there is no need for bumper blocks which are used chiefly in case of delicate perforating and/or bottoming forming punches, such as coining.

Second Operation Press Tools

One must dedicate more care to second operation dies because they are far more dangerous than first operation tools. Statistics show that for every five punch press accidents, four happen with second operation dies. The reason for this is that for such tools there is, unfortunately, an actual need for reaching with the hands "under-the-punch," i.e., into danger zone (1), in order to insert or remove material. (Exceptions are favorably designed—usually large—workpieces where only a part of the surface is involved in the operation. In such cases, the stampings may sometimes be held by gripping them outside of the die, far away from the nip point.) Therefore, the first task for the tool designer consists of finding some artificial means by which the need for putting the fingers into the area of operation [danger zone (1)] is eliminated. Such artificial means are suitable, safe feeding and ejecting methods and devices. Once such means are incorporated into the design of the tool, proper protective guards (exactly like those for first operation dies) can be provided for screening the danger area, in order to prevent the possibility of inadvertently reaching into it, and thus the goal of safe working conditions is achieved.

Feeding in Second Operation Tools

In case of first operation tools, the method of feeding the stock had little or no influence on the operator's safety. With a correctly protected tool, practically any kind of feeding may be used safely. Vice versa on the other hand, in the case of second operation tools, the correct and adequate feeding method affects directly the operator's safety.

As was mentioned in previous chapters, in any kind of press operation, one of the following three feeding methods may be employed: manual, semiautomatic, or fully automatic. In the manual feeding, the material to be processed is put under the punch by hand, directly or indirectly (i.e., with some device—tool or holder). In semiautomatic feeding, the material is placed under the punch by some mechanical or other device that requires the direct intervention of an operator at each press stroke. (The operator's intervention consists of handling the material—at a safe distance, far away from the danger zones—and of initiating the press cycling.) Fully automatic feeding does not require the constant attention of an operator at each press stroke; his only feeding activity consists of filling periodically some magazine or other feeding attachment with blanks or stock to be processed. The material is then placed under the punch automatically by a mechanically actuated appliance, for instance, and the press is cycled automatically, without interruption, at a rhythm dictated by the press.

From both the safety and the production efficiency standpoints, of course, the fully automatic feeding method is the best. Next in line is semiautomatic feeding and then manual feeding, which should always be made *indirectly;* direct manual feeding is highly dangerous and should be prohibited.

Ejection in Second Operation Dies

Obviously, the removal of the finished stampings and of the scrap material from the die must also be performed safely, without the need for the operator to reach with his fingers into the interior of the tool. Consequently, if the evacuation of the workpiece is not done automatically through the proper clearance holes of the press table, as happens with blanks and with flangeless drawn shells and other analogously formed stampings, the ejection is performed in two consecutive steps: first the workpiece is stripped, i.e., separated from the punch and/or the die, and then it is removed by some adequate means, preferably toward the rear of the press, where it is collected in proper tote boxes or bins.

The stripping is done in the usual way: In cutting dies, with stripper plates, in forming dies, both with stripper plates (if needed) and with push-outs or kick-outs (actuated by springs or rubber pads, mechanical devices, pneumatic cylinders, etc.; see Chapter 11) located in the female stamps. Sometimes, also, suction breaker push-off pins are employed for separating cut and/or formed stampings from the respective active tool members. The actual ejection may be automatic or manual. Press tilting, compressed air jets, mechanical kick-outs, etc., are used for the first method; pushing slides, rods, or plates and gripping devices are used for the manual ejection. In any case, the manual ejection must be always *indirect.*

In the case of second operation dies, it is recommended that special attention be paid to ejection, since more accidents happen during part removal than during feeding.

Scrap Removal

The die designer must also pay attention to the safe removal of waste produced in the operation of a die. This removal should, if possible, always be automatic, without any necessity for the operator to put his hands into the danger zones of the die. The solutions are analogous to part unloading: gravity, air blasts, chutes, grippers, conveyors, etc., are utilized for this purpose. Automatic scrap cutters are very useful; they cut strip skeletons into small pieces that are easier to handle.

Positive Acting Safety Devices

There are cases where it is not convenient (economically or technically impractical) to install point of operation guards. Then positive acting safety devices should be provided which offer some protection. Several of these utilize the downward movement of the press-ram to actuate a mechanism that prevents the operator from putting his hands into the danger zone when the machine is operating. Among these safety devices are the following principal types:

a. Interlocked gate guards that consist essentially of movable (telescoping or pivoting) adjustable transparent gates or cages, which cover the feeding opening when the press ram is in motion. (Sometimes, for greater visibility, the front part of the gate is made of a sheet of transparent plastic material, so that the operator is better able to check the performance of the operation cycle.) There is a mechanical (or some other type) interlock which makes it impossible to put the ram in motion when the gate guard is open, even if the latter is stopped in any intermediate position by a hand or any object, and once the ram starts to move, it is not possible to open the guard gate until the ram is in its top, rest position.

The closing of the gate may be performed either automatically or manually. Automatic return to open position is always effected without need of the operator's intervention; it is done with springs, balance weights, or other suitable means. In sequential operations, it is the gate that automatically starts the clutch engagement precisely when the gate is completely closed (total safe covering of the danger zone is achieved).

With interlocked gate guards, if the press is equipped with a positive-type clutch, this is usually combined also with the "nonrepeat" (single-stroke) mechanism of the press. It limits the travel of the ram to one complete cycle at each tripping of the clutch. This means that if the operator, because of a distraction or attempting to remove a workpiece, keeps holding his foot on the clutch pedal, second strokes are prevented; the ram remains in the top, rest position after one single stroke; in order to start a new ram stroke, the pedal must be released and then reactuated. This device greatly improves the accident prevention conditions. By the way, "nonrepeat" devices can be retrofitted to existing presses.

b. Pull-out guards are rather efficient devices, but they are not very popular, because of their allegedly complicated working. The operator wears a pair of wrist or elbow straps made of leather, or other suitable material, provided with rings where snap hooks are fastened. These hooks are on the ends of two small ropes or chains which, united, form one single cable. The latter passes behind the operator, through a pulley located at the top of a post, which is attached to the floor or to the press frame, and from there to a multiplier mechanism which in turn is connected with the press-ram, plunger, or outer slide. When this latter begins to descend, the cables pull the operator's hands until they are completely withdrawn from the danger zone of the die.

c. Sweep (pendulum) guards consist of one, or occasionally two, vertically or horizontally oscillating arm(s), actuated by some movable part of the press, usually the ram, through a proper articulated mechanism. When the press ram starts its descent, the sweeping arm is moved mechanically through the front side of the press so that if the operator's hand is in the danger zone, it is pushed aside gently yet firmly. The arm is usually made in the form of a metal rod and its end, which comes into contact with the operator's hand or arm, is properly padded with leather, rubber, plastic, or some other suitable, soft material. There must also be a flag or barrier so that the operator cannot reach behind the sweep while it is in motion.

d. Two-hand tripping controls (starting release) are indirect protective devices. They oblige the operator to keep *both* hands well outside of every possible danger zone, in order to start the press ram movement. The clutch may be engaged and thus the ram started, only by simultaneously pushing two actuating buttons or levers. There are several practical solutions for the actual realization of this task. They may work mechanically, electrically, electronically, or pneumatically. For presses that are operated simultaneously by more than one person if one set of two actuating buttons or levers is provided for each person or each operation station, there is complete protection for all of them. The buttons should be shielded and conveniently located (e.g., head height) so that they cannot be operated with any part of the operator's body except with his hands, as intended. Also, they must be located far enough from each other that the operator is obliged to use one hand for each of them. It is strongly recommended to use a "nonrepeat" attachment for the press with these two-hand tripping devices, as employed for interlocked gate guards.

e. Photoelectric-type guards (Figs. 13-18 to 13-20) eliminate the use of physical barriers. They provide a source of invisible light (emitter or transmitter) that produces a "curtain" covering the front of the press; if necessary, the sides and the rear of the press may also be protected. The light is received by a corresponding sensing electronic device properly connected to the clutch brake of the press. The press may be operated only if the light curtain is not interrupted by the operator's hands or body or any other object. In addition, if the light curtain is broken while the ram is already in motion, the press is stopped instantly—the friction clutch is disengaged and the brake applied automatically.

Electronic safety devices should not be used on presses having positive clutches ("full revolution clutches," i.e., where the ram continues a complete stroke after each tripping of the press).

Note: In addition to some of the positive-acting safety devices, special gripping tools may be used as a supplementary safety means. (In accident prevention any-

Fig. 13-18. Accident prevention through invisible light curtain. (*Courtesy of Weldotron Corporation*)

Fig. 13-19. Accident prevention through invisible light curtain. (*Courtesy of Data Instrument Inc.*)

thing that may help should be welcome.) The same kind of gripping hand tools are used as are employed for feeding and unloading in second operation tools. See Figs. 4-4 and 13-18 where safety tongs are attached permanently to the presses.

Limitations of Positive-Acting Safety Devices

These appliances are practically always commercially produced by specialized manufacturers. It is claimed by them that their products represent effective protection. This may be true in the overwhelming majority of cases; however, it is not always so. Official and private reports indicate that—in exceptional cases—accidents have happened even when presses were provided with such protective devices.

Notwithstanding this, positive-acting protective devices must be taken into consideration; they have their own well-defined place. They must be used in those cases of very dangerous operation where the tool simply cannot be provided with combi-

Fig. 13-20. Accident prevention through invisible light curtain. (*Courtesy of Data Instrument Inc.*)

nation guards, because the operation demands accessibility to the press by workers who must reach into the point of operation. They are also useful for short run jobs. In all such cases, positive-acting safety devices are better than nothing.

Protection for the Presses

In order to achieve complete safety for the workers and for other people who occasionally or frequently are in a metal stamping shop, it is necessary to provide protection guards against accidental contact, not only for the immediate surroundings of the die, but also for the whole machine and its attachments. Therefore, for every moving (rotating or sliding) part, viz., flywheel, pulleys, belts, gears, covering protectors must be provided, made in the form of solid or transparent metal boxes as is done with any kind of production machinery. Not only the press and its parts, but also its attachments—e.g., feeding and unloading devices—must be properly protected.

It is recommended that the protective guards for presses be made in removable sections, preferably hinged, so that, when necessary, it would be easy to move the flywheel manually to inspect, service, and repair the press. Mechanical presses with overhanging flywheels, where there is no external supporting bearing beyond the flywheel, should have well constructed guards capable of holding the flywheel in case of shaft breakage.

In case of presses actuated by foot pedal, a strong guard must be provided above the pedal, which would prevent the accidental triggering of the press (see Fig. 13-18). Sometimes, special pedals of intricate designs are employed for the same purpose.

Safety in Material Handling

Die handling: In order to avoid accidents due to heavy dies which if dropped may injure die setters and their helpers, even during the design stage of the tools, their proper handling must be taken into consideration. Eyebolts (in the die holder) grooves, must be provided for lifting the die with lift trucks, cranes, and/or other adequate mechanical lifting devices.

It is recommended that one should round or chamfer every external sharp edge of dies, in order to avoid hand and finger injuries during their handling.

During the handling of dies a few special precautions must be observed, such as: avoid excessive muscular strains, always utilize mechanical lifting means if the tools are heavy. Look out for sharp edges. Look out for your fellow workers. Wear safety shoes, gloves, and correct clothing.

Material and scrap handling: material and scrap must also be handled with proper care in order to avoid accidents and problems. The persons assigned to this type of work should follow proper safety rules, such as: Wear proper clothing (heavy apron, safety shoes, gloves, arm protectors, etc.). Use adequate tools and take care that they are always in excellent working condition. Hooks, rakes, forks, and other material-handling aids have to be used correctly and carefully.

Guard Maintenance

According to type, design, construction, and use, guards require more or less care. The simple, solid, fixed combination guard (without moving parts) requires practically no maintenance; universal guards, periodical slight inspection; and positive-acting devices, much more. For example, with interlocked gate guards a very high standard of maintenance is indispensable, because a mechanical failure of the interlocking mechanism may completely impair the safety efficiency without the operator being aware of it, leaving him with a false sense of security. Thorough examination and careful inspection at each fresh setting and resetting of tools is needed. With long production runs, guards should be tested at the beginning of each working period. Practically all other positive-acting protective devices need some kind of very careful maintenance and inspection.

Practical Hints

There are a few additional aids of a technical character for safety:

a. Install additional stop buttons on every press, so that there is one on the right and one on the left side uprights, and at the front and at the back of the press. Then, in case of emergencies anyone approaching the press from any direction can stop the machine without fumbling around and wasting precious seconds.

b. Paint in white the inner surfaces of the press frames. In this way, the operator is better able to distinguish between the work and the die, consequently, he works more safely, surely, and quickly.

Conclusions

Of course, adequate tool design and guards *alone*, are not sufficient to achieve effective, complete accident prevention. They must be supplemented and supported with anything that may help. Among the supplementary aids the following items should be mentioned:

Agreeable and pleasant working conditions in the plant

Preventive press and guard maintenance

Good housekeeping

Education and training of personnel for safety

Cooperation from outside agencies; information exchange

Accident prevention adds a fringe benefit of ecnomic advantages, which are of considerable magnitude. Apart from the negative circumstance that the direct and hidden costs of accidents are avoided, there are also positive, tangible results such as increased production, improved quality of products, decreased spoilage, increased tool life, and improved employee morale.

CHAPTER FOURTEEN

Protecting Equipment from Damage

Accident prevention activity extends beyond the protection of people alone. Precautions and safety devices must also be provided in the press shop for protecting tools, presses, and other pieces of equipment. This activity has two goals: additional means to prevent accidents to the personnel and to prevent damage to the production equipment.

Damage to equipment reduces productivity in many ways. First, there are production interruptions. Then, the tools must be removed from the press and after repair, set up again; this means extra diesetting labor. Finally, the repair job costs money in toolmaker's labor, spare parts for replacement, etc. Furthermore, whenever an operating tool is damaged, there is frequently an accident hazard for anyone nearby. During a mechanical "short," quite often some hardened tool member is broken and small pieces of hard steel fly around like bullets. Any operator, diesetter, floorman, or other person who is within the range of these pieces may be injured. Therefore, in many cases the tools and/or the presses are equipped with special features or with extra protective devices. The purpose of such an arrangement is this: In case of any irregular behavior in the operation of the tool, the press must be stopped, possibly before any serious damage or accident can develop.

This chapter is divided into two separate parts. The first covers press-tool designs that incorporate the means to prevent *tool malfunction* and in this way to avoid rejects, accidents, press shutdowns, or other kinds of problems. The second part covers devices that provide protection against the consequences of press and press-tool *overloads.*

Safety of Press Tools

The goal of die-accident-prevention devices consists of (a) making it impossible to trip the press (in case of single-stroke operation) or (b) stopping the press (in case of continuously running operation).

The first group comprises monitors of such irregularities as improper registering of tool members, faulty nesting or positioning of blanks in second operation dies, failure to eject finished stampings, slug pulling, double feeding, stock thickness out of tolerance limits, etc.

To the second group belong devices which monitor the following items; misfeeds (overfeed or underfeed), stock loop (buckling), pile-ups, end of coil, stock width (too wide or too narrow), etc. Some devices are used both for first and second operation tools.

The practical realization of the monitoring details or devices, in general, is quite inexpensive. At any rate, according to general consensus, it is considered that such expenses usually are a small fraction of the savings realized through the avoidance of just one die smash-up caused by some malfunction.

The usual means employed for tool safety devices are simple or sophisticated mechanical, pneumatic, hydraulic, electrical, electromechanical, electronic, and photoelectric, employing limit switches, microswitches, probes and sensors of many kinds, missing pulse detectors, missing part detectors, proximity coils, light curtains, and impact plates. The accompanying sketches (Figs. 14-1 to 14-20) present a small selection of typical applications, taken from many actual cases.

Improper registering of tool members: Figure 14-1 shows one of the most common and simplest cases: It is the monitoring of a reciprocating slide feeder for secondary punching operation. A microswitch is located so that the press cannot be tripped until the slide is pushed into position with the part to be processed lined up correctly beneath the punch.

The case of Fig. 14-2 is similar to that of Fig. 14-1 with the difference that here it is the whole die plate which is movable and is mounted on the slide. Observe the protecting tubing shield around the actuating pin of the microswitch. This shield prevents accidental tripping of the microswitch.

Faulty nesting of blanks: In the case of simultaneous processing of several workpieces in a given tool, each of the part nests must be controlled with a separate switch, and all the switches are connected in series in the press-tripping circuit (Fig. 14-3). Unless all the nests are loaded correctly, the press won't trip.

Failure of ejection: For the protection of a double-acting operation (both cutting and drawing), the knock-out bar is controlled with a microswitch (Fig. 14-4).

Figure 14-5 represents a combination die where the blank is cut off and bent. A limit switch controls the position of the ejector pad. If this is stuck at the bottom position (i.e., it does not go back up to its normal position), the press cannot be tripped for the next operation cycle.

In Fig. 14-6 the drawn shells are evacuated by an air jet. If the part is not unloaded (because the ejector pad is jammed or the air jet failed to work, or whatever reason), the microswitch is not actuated and the press cannot be tripped again.

In Fig. 14-7 stampings are passed through a magnetic field (proximity sensor) which must be altered momentarily by the steel part in order to allow the press to be tripped.

Slug pulling: For preventing the consequences of slug pulling, one of the simplest solutions (Fig. 14-8) consists of prolonging the oil film breaker pin, letting the top end protrude so that it actuates a NO microswitch. The press can be tripped only if the pin has separated the slug from the punch bottom.

Misfeeds: In blanking dies the advance stop pin can be made swinging (Fig. 14-9) and spring-loaded. When the strip skeleton is correctly advanced and the last blanking opening in the strip moves the stop pin into the "go" position, a NO microswitch is actuated so the press can be tripped.

In case of French stops (Fig. 14-10) the trimmed coil piece is pushed against a pivoting lever which has a double function: one end is the advance stop, the other end is the microswitch actuator.

In the case of cut-off dies, the far end of the coil must actuate a microswitch in order to allow the tripping of the press. In case of stiff stock, 0.080 inch and over (Fig. 14-11), a direct plunger guided in a steel bushing block is pushed against the micro-

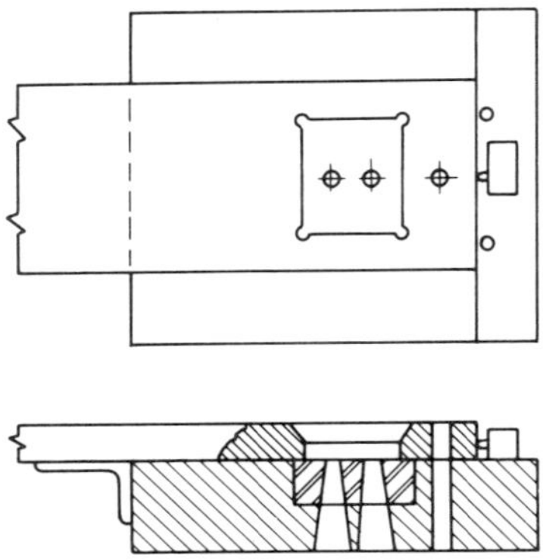

Fig. 14-1. Microswitch actuated by improper registering of feeding slide.

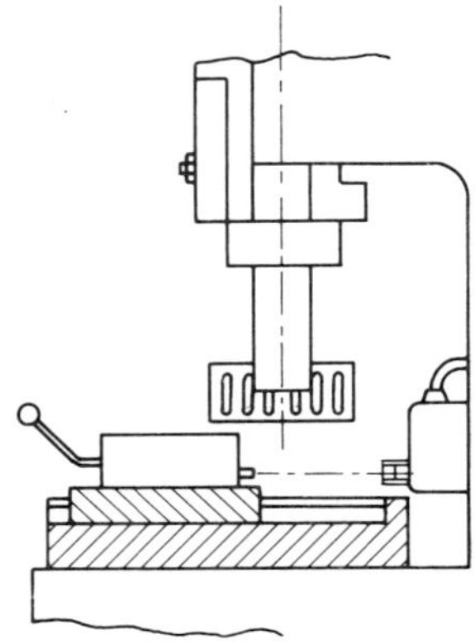

Fig. 14-2. Microswitch actuated by incorrect positioning of reciprocating die.

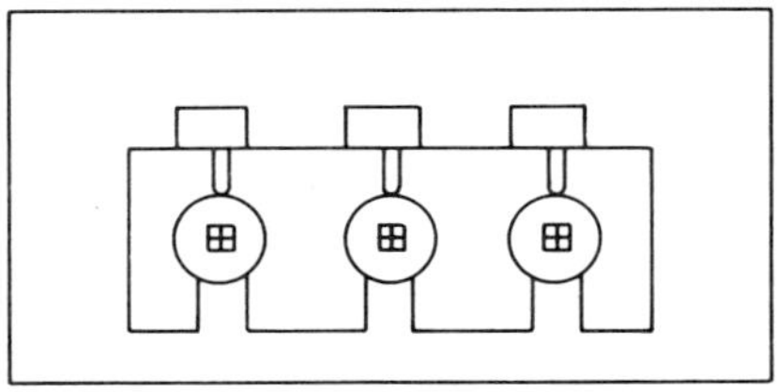

Fig. 14-3. Multiple checking with microswitches.

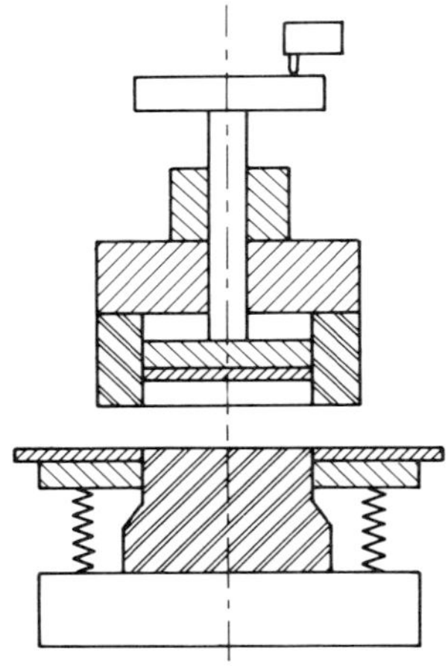

Fig. 14-4. Microswitch indicates failure of ejection in compound die.

switch pin. In the case of medium gage stock, 0.048–0.080 inch (Fig. 14-12), a pivoting member is intermediary between coil end and switch pin. In the case of light stock, under 0.048 inch (Fig. 14-13), a low-force switch rod actuator is moved by a swinging angle arm part.

Strip-width monitor: A bellcrank pivoted between two switches acts to magnify excessive deviation of the strip from maximum or minimum tolerances when use of "exact width" is essential (Fig. 14-14).

Buckle detector: Sometimes a strip will buckle or kink and cause problems in dies. When the defect comes along, the switch is tripped to stop the operation (Fig. 14-15).

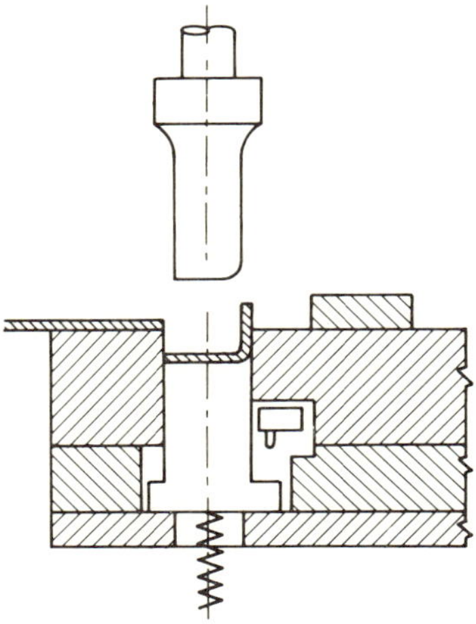

Fig. 14-5. Ejector pad checked by microswitch.

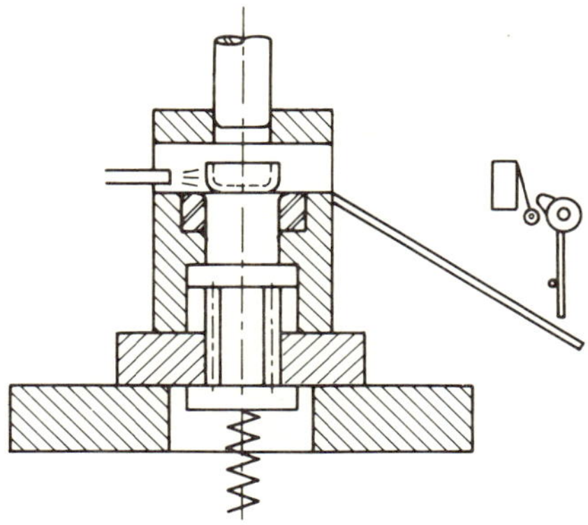

Fig. 14-6. Microswitch indicates failure of ejection of drawn shells.

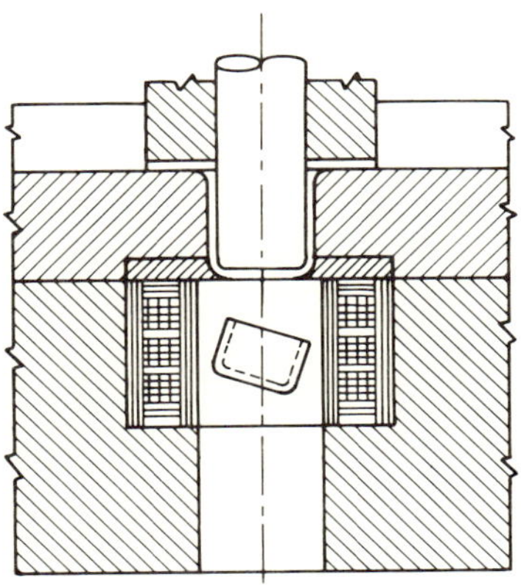

Fig. 14-7. Magnetic proximity sensor for part ejection.

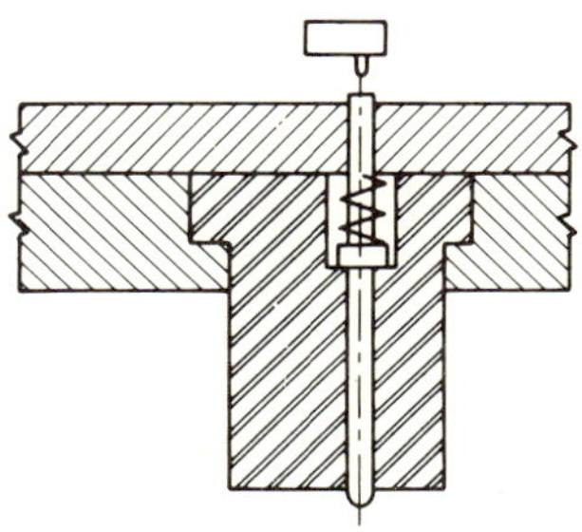

Fig. 14-8. Oil film breaker actuates microswitch.

Strip protector: If something adheres to, or is being carried along by the strip, this switch will stop the press and feed (Fig. 14-16).

Double-blank detector: Two blanks or an extra-thick piece of stock in the nest of a second operation die will stop the press (Fig. 14-17).

Coil end detector: When the coil end leaves the switch the scrap recoiler is stopped so that the free end does not fall into the die (Fig. 14-18).

Stack monitor: The press is stopped when the magazine feed runs out of blanks (Fig. 14-19).

Figure 14-20 shows the control panel of a press control monitor of malfunctions, as detailed above.

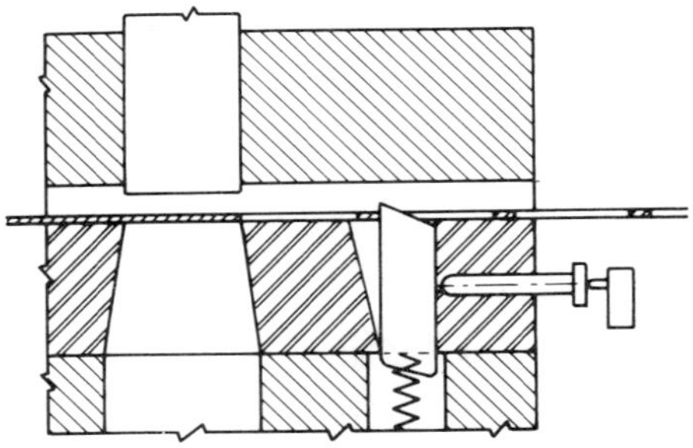

Fig. 14-9. Advance stop is simultaneously a probe, actuating a microswitch.

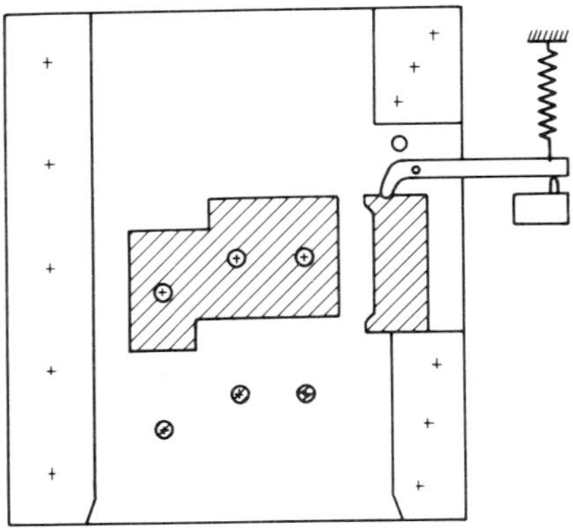

Fig. 14-10. French stop monitor.

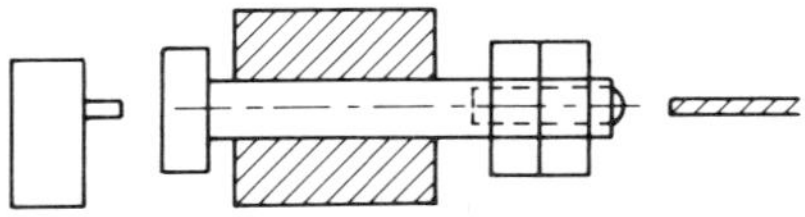

Fig. 14-11. Cut-off die monitor for heavy stock.

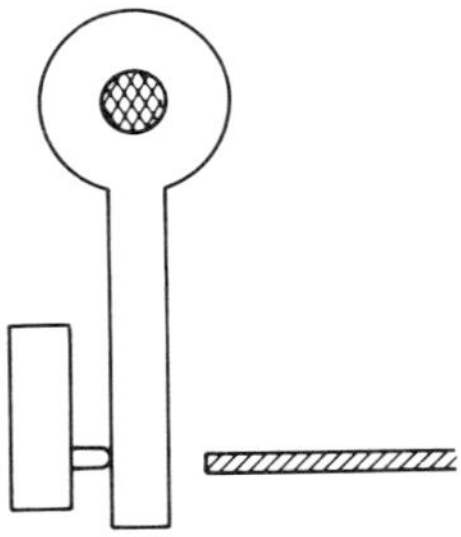

Fig. 14-12. Cut-off die monitor for medium gage stock.

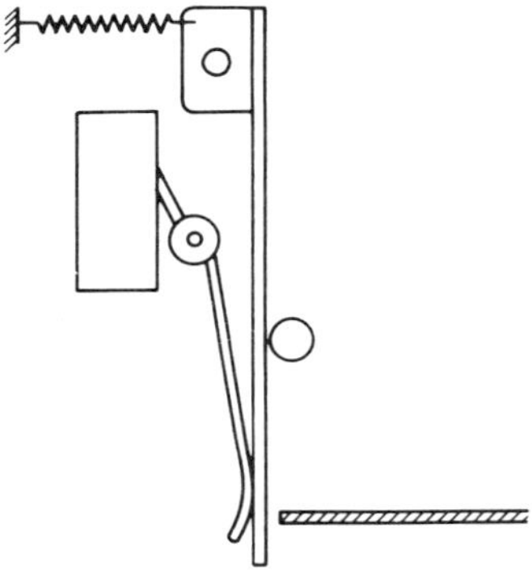

Fig. 14-13. Cut-off die monitor for thin stock.

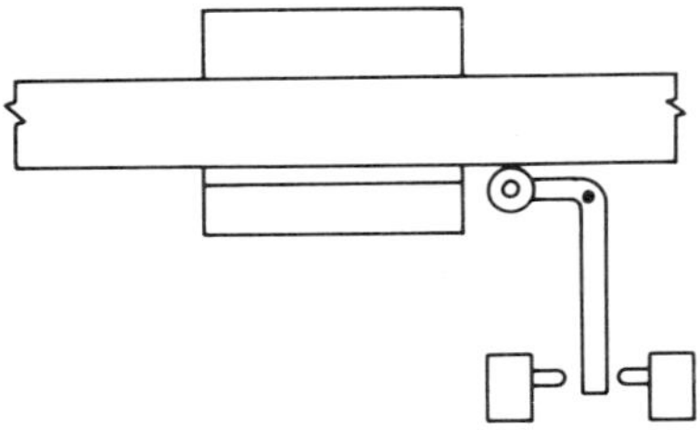

Fig. 14-14. Strip-width monitor.

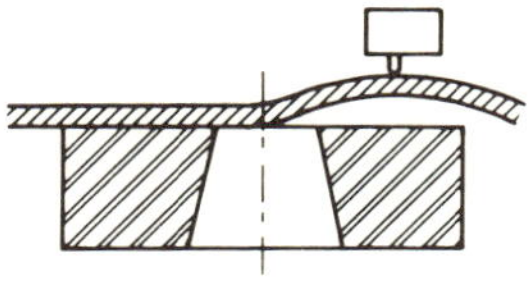

Fig. 14-15. Buckle detector.

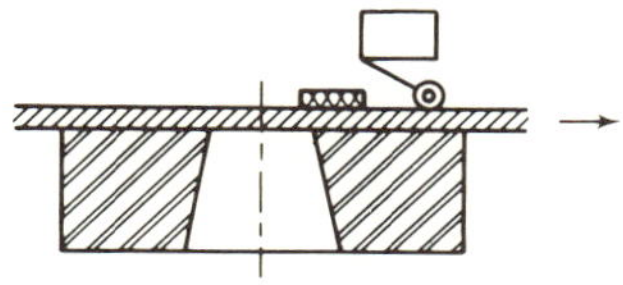

Fig. 14-16. Strip protector.

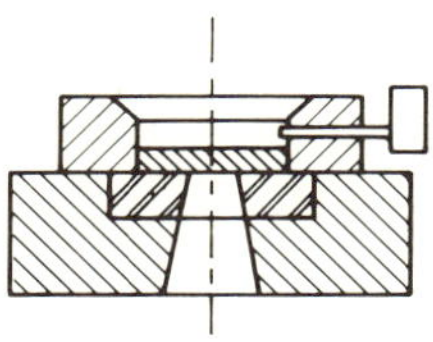

Fig. 14-17. Double blank detector.

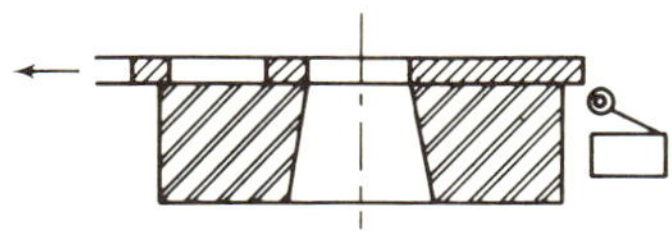

Fig. 14-18. Coil end monitor.

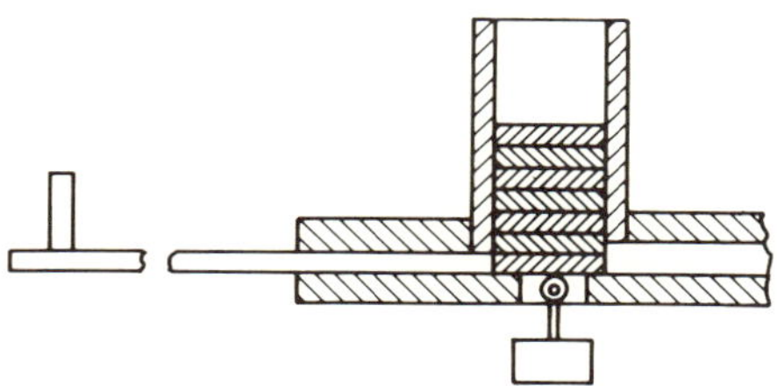

Fig. 14-19. Stock supply monitor for second operation die.

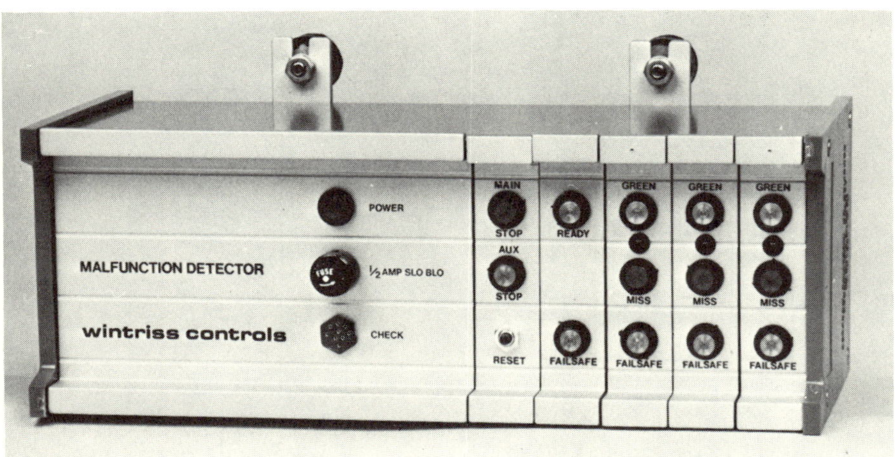

Fig. 14-20. Press malfunction panel. (*Courtesy of Data Instruments Inc.*)

Press Protection Against Overloads

Methods for preventing damage to press tools by overloads have gone through a basic transformation during the last few decades. Formerly—and in many cases of simple tools, and especially in small companies, still today—for the protection against overloads of dies only mechanical and/or electrical, rarely hydraulic or pneumatic, devices were employed. Figures 14-21 to 14-31 present several arrange-

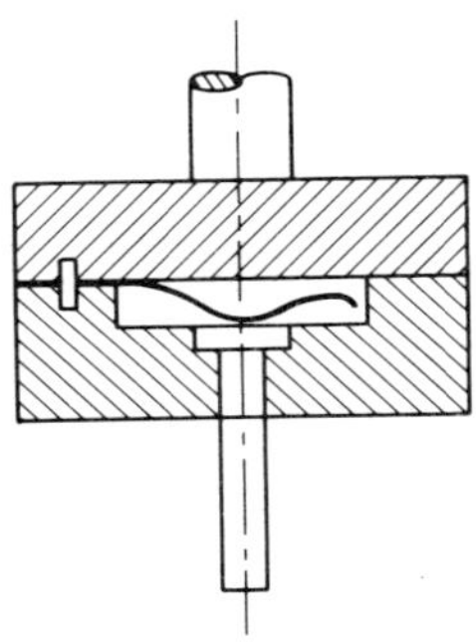

Fig. 14-21. Flat spring as overload protector.

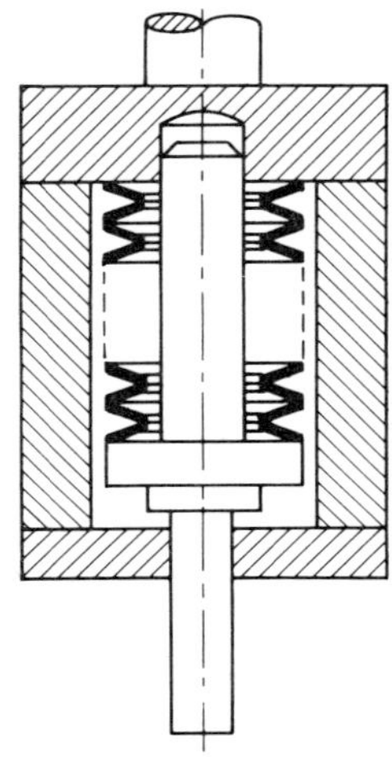

Fig. 14-22. Washer springs as overload protector.

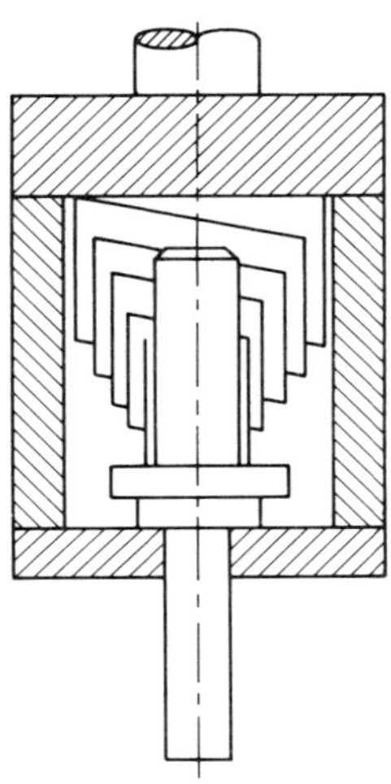

Fig. 14-23. Taper springs as overload protector.

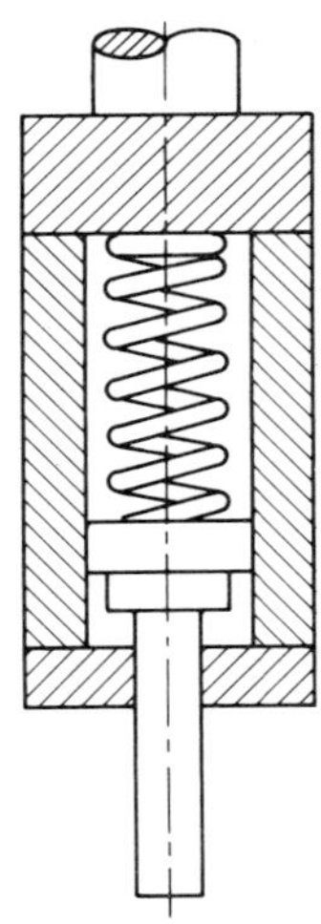

Fig. 14-24. Helicoidal spring as overload protector.

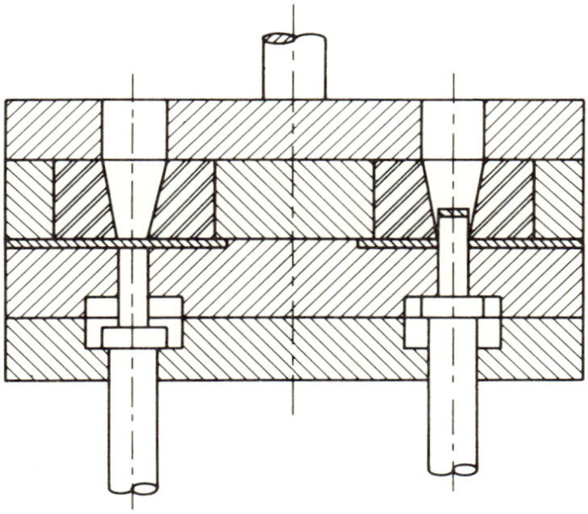

Fig. 14-25. Shear plates are perforated in case of overloads.

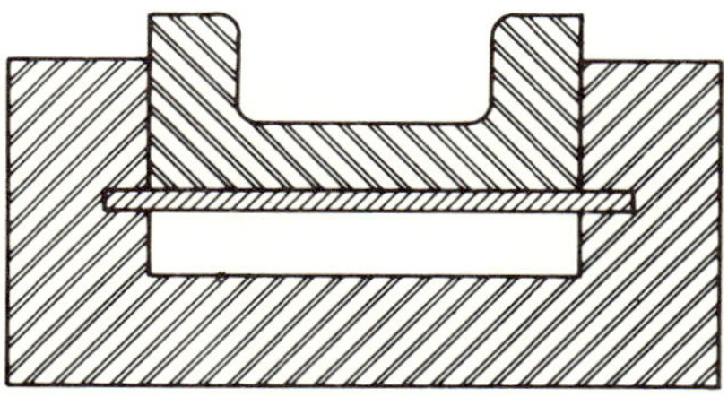

Fig. 14-26. In this forming die the shear plate is severed in case of overload.

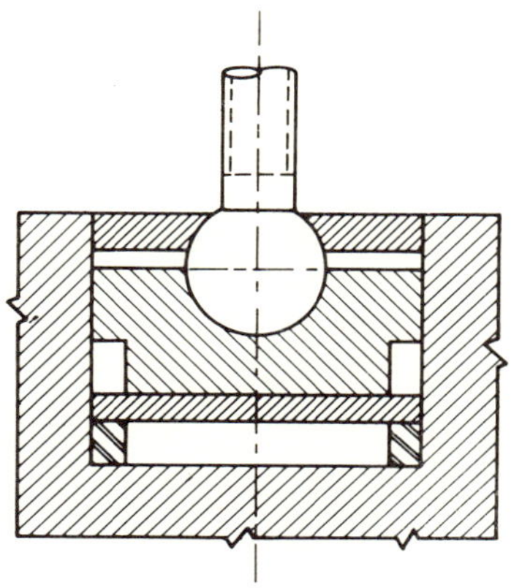

Fig. 14-27. Shear plate mounted in top of ram for press protection.

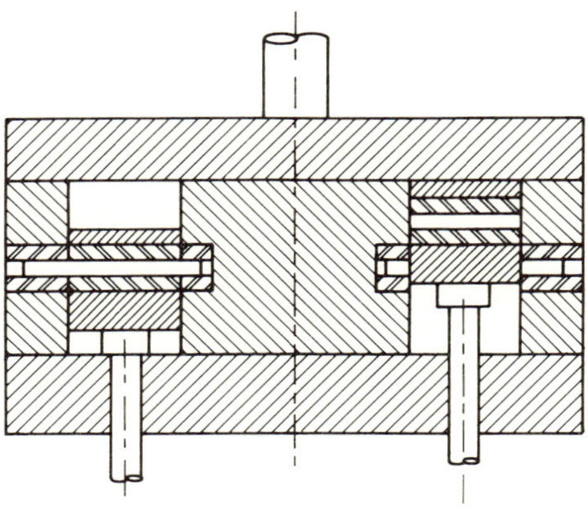

Fig. 14-28. Shear pin is sheared in case of overload.

Fig. 14-29. Wintriss/Load Analyzer. (*Courtesy of Data Instruments Inc.*)

Fig. 14-30. Overload monitor control panel. (*Courtesy of Weldotron Corporation*)

Fig. 14-31. Overload monitor installed on a press. (*Courtesy of Toledo Transducers International Sales Corp.*)

ments for preventing damage to press tools in case of overloads or overstresses caused by double blanking, double punching, double forming, inadequate stock, tool member disalignment, trouble with slugs, lost lubrication, and other possible problems.

Mechanical Devices

The mechanical overload protecting devices presented here may be classified into two groups (a) permanent, i.e. springs, and (b) replaceable ("safety fuses"), i.e., shear plates or shear pins.

In all cases the safety element is placed in a specially designed punch holder. Obviously, for replaceable members, the tool must be so designed and built that the "safety fuse" element is changed easily and quickly, just like an electrical fuse.

The safety device is so calculated that it will resist the normal, nominal load and a small overload up to approximately 25–30%. For overloads greater than this limit, they react by rupturing or deflecting (yielding).

Experience has shown that "safety fuse" elements are best made from comparatively hard stock, because this needs less penetration than soft stock before it is cut through.

Springs: Several kinds of springs may be employed for overload protection of piercing punches, such as: (1) flat springs (Fig. 14-21), (2) Belleville (washer) springs (Fig. 14-22), (3) taper springs (Fig. 14-23), (4) ordinary helicoidal springs (Fig. 14-24), (5) ring springs, (6) rubber springs, (7) rubber cushions.

Note: In the designs where springs are acting directly, (Figs. 14-21 to 14-24), the protection devices are reset automatically after removal of the overload, i.e., after the press ram has reached the top, rest position.

Shear-plates: Figure 14-25 illustrates the working principle of shear plates for piercing punches and Fig. 14-26 for forming dies. Analogous to the last design is the one that protects not only the tool but also the press. It is shown diagrammatically in Fig. 14-27. It is mounted in the top of the press ram. The shear plate is cut through when the press load exceeds the preestablished limits.

In case of piercing punches in another design, the shear plates are located between the prolongation shaft of the special piercing punch and a matching die button.

Shear pins: In Fig. 14-28 is shown the application of shear pins. These are usually made from unhardened drill rod and are loaded in hardened bushings put into both the punch holder and the auxiliary filler block.

Automatic Stopping of the Press

It is very sound practice, which should be followed as a rule, to stop the press automatically whenever the overload protection device is actuated. Even in those cases where the resetting is automatic (spring-actuated devices) the press operator must be made aware of the fact that something went wrong. In case of fully automatic production (presses without a constant attendant operator or at least without a directly intervening operator) there should also be an optical (and possibly also an acoustical) signal to warn the people (press attendant, foreman, etc.) about the occurrence.

Electrical Devices

Today the tendency is for more sophisticated electronic monitoring devices that actuate indirectly, stopping the press (disconnecting the driving motor and applying the brake to the press) in case of overloads.

The majority of overload monitoring systems are so arranged that they take care not only of the protection against press and tool overloads, but also of the majority of die malfunctions. A typical example of such a device is the Wintriss Load Analyzer (see Fig. 14-29), which is a modular electronic load measuring and monitoring device that checks constantly the different characteristics of press operations, preventing possible die and press damage caused by overloads and/or malfunctions.

Figure 14-30 presents analogous control panel of Weldotron Corporation. Finally, Fig. 14-31 shows the actual installation of a monitor made by Toledo Transducers.

CHAPTER FIFTEEN

Troubleshooting

Problems with Cutting Die Plates

As stated before repeatedly, any unscheduled or forced production interruption means a decrease in productivity. Therefore, every detail must be taken into consideration that may result in unplanned downtime.

Die plates of cutting dies may cause trouble by: (1) dulling of cutting edges and consequent excessive burr formation or (2) breakage.

Cutting Die Plate Performance

The useful life of a cutting die (blanking, punching, trimming, notching, etc.) is determined simply by the total amount of satisfactory stampings produced with the die.

In operation, cutting edges of the tool must withstand the cutting strain due to the stock resistance and the considerable wear caused by friction. After a certain number of operations the keenness of the cutting edges is lost, they become rounded and dull. Consequently, the burrs on the outer contour of blanks and/or around the periphery of punched holes become excessive. (The admissible maximum burr height may depend on the part's function or the assembly method or aesthetics or whatever; see more details in next item.) In order to restore the original keenness to the cutting edges, the die plate and the punch are reconditioned by a slight regrinding. In the case of high-production tools, this resharpening operation is repeated at intervals until finally the tool fails to produce satisfactory stampings due to enlargement of the die opening or the tool breaks because of insufficient strength.

Determining Factors of Tool Life

How wear and consequent cutting edge dulling are created is not fully understood in every detail, because there are too many factors which affect them and these factors cannot be separately and independently examined and evaluated. The determining factors may be classified broadly in four groups, depending on: (1) the stamping; (2) the tool; (3) the press; and (4) the operation.

We considered here only those factors that belong to groups (3) and (4). Those of group (1) are determined by the component or product designer, and those of group (2) depend on the tool design. Therefore, both of these groups are outside the scope of this book.

To the press group belong the following chief items: press selection, cutting speed, and tool setting.

Press Selection

For any kind of die the selection of a good press of the right capacity and maintained in proper working condition is of extreme importance from the productivity standpoint. Every C-frame press (which is the type most commonly used) at the instant of cutting impact, deflects somewhat, misaligning the punch a little. If the press capacity is very high with respect to the force required by the cutting action of the die, then the press deflection is very small, almost insignificant, and no harm is done; on the other hand, if the deflection is too pronunced, then the misalignment becomes too great so that uneven wear is produced, and in extreme cases even damage to the cutting edges occurs. Consequently, it is recommended that, where feasible, a much larger press be selected than is strictly necessary in order to prevent undue press frame deflections. In addition, if there is a possibility of choosing between an OBI press and a straight side press, take the latter one. It deflects much less and so ensures longer tool life.

Cutting Speed

About the cutting speed, expressed in feet per second at the instant the punch contacts the stock surface, different sources report rather consistent information. Everyone advocates high speeds, within reason. Repeated laboratory experiments, corroborated by checks in actual practice, have demonstrated that the temperatures reached during high-speed cutting processes do not impair the hardness of the cutting edges.

In addition, it seems that high speed improves the conditions that prevent premature burr formation. For example, according to some British experts, in the case of ultra-high-speed production presses—1000 strokes per minute or more—clean cuts are produced even with what might otherwise be considered as excessive punch-to-die clearances.

Die Setting

The craftsmanship, skill, and care with which the die setter performs his job play an important role in tool performance. The proper setting of the tool in the press is of utmost importance, not only for open-type dies but also for those tools that are provided with die sets. In fact, if the alignment is not correct, excessive friction and certain dangerous lateral stresses are produced which cause premature wear of the tool members.

To the operation group belong the following chief items: operation proper, lubrication of stock, and lubrication of the tool.

Operation with the Tool

The press shop contributes in several ways to tool performance. One of the first requirements consists of the periodic cleaning of the tool in order to avoid accumulation of dirt, chips, or foreign matter, which, mixed with the lubricant, act as abrasives and considerably increase tool wear.

The press operator is supposed to cooperate in the prolongation of the useful life of the tool by: (a) avoiding any irregularities (double blanks put into the nest, faulty nesting of blanks, inaccurate feeding); (b) informing the foreman immediately when burrs start to become excessive; (c) in general, giving the tool the care it deserves as an expensive precision product.

Proper lubrication and cooling of the metal to be stamped are great helps in facilitating the stamping operation. The direct consequence is—among other benefits—a notable prolongation of tool life, because the friction between the workpieces (blanks or slugs) and cutting members of the die is greatly reduced.

Also, the moving members of the tool must be properly lubricated just like any moving part of a machine. Therefore, the bushings and leader pins of the die sets are to be greased constantly, in order to avoid galling.

Practical Hints

Here follow a few practical hints with respect to prolongation of the useful life of cutting dies.

Sharpen the Die More Often

There is a very simple, inexpensive method which makes it possible to prolong notably the useful life of the cutting members of press tools. It is a detail generally overlooked by toolmakers and press shop people.

The curve in Fig. 15-1 shows what happens normally to burr height with the change in cutting edge keenness of dies. With a new or freshly resharpened tool, at the start, there is very little burr (it is common practice to evaluate the cutting edge conditions through the burr height on the stamping edges or around the hole contours). However, very shortly, the cutting edges become more or less slightly rounded and a small burr formation starts at the blank and/or hole peripheries. After a rather limited number of cuts, the rounding off of the cutting edges becomes stabilized (point A), and the burr height increases from this point only imperceptibly; the curve becomes almost horizontal. This is the normal part of the run. After a given quantity of hits and rounding of the cutting edges becomes suddenly larger (point B), and the curve turns steeply upward at about 45°. This is the point when the operation with the tool is stopped by the inspector, and the tool must be resharpened.

The moment (i.e., the quantity of hits made with the tool when the die must be reconditioned by grinding does not necessarily coincide with the burr reaching the maximum allowed height. In many cases, smart shop people do not wait, but stop the

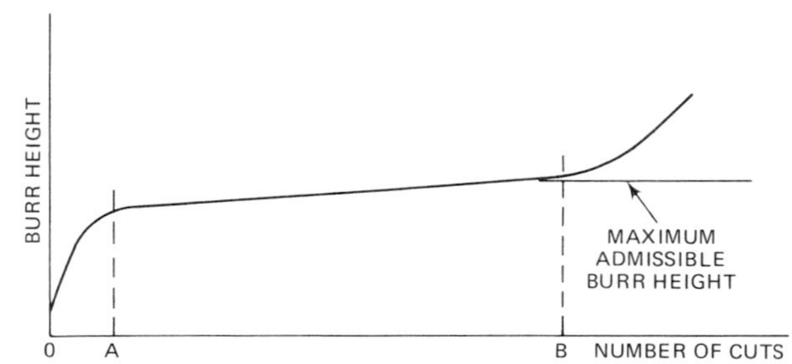

Fig. 15-1. Normal curve of burr height showing increase with number of cutting die operations.

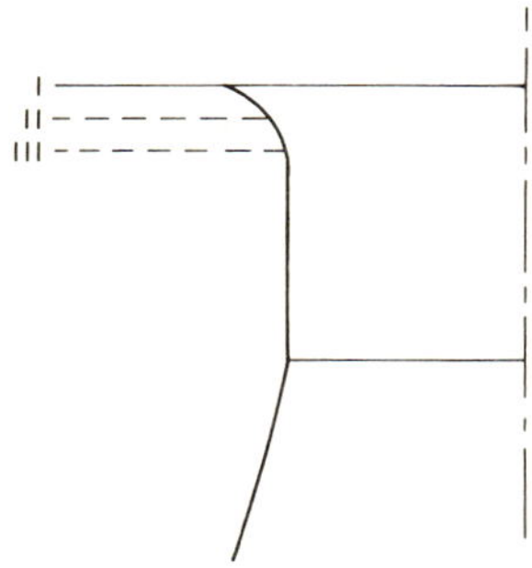

Fig. 15-2. Suggested sharpening levels for die plates and punches.

operation just before the curve would change direction (before reaching point B), even if the burr height is still less than the allowed maximum limit. An additional suggestion for getting more out of a cutting die deals with the first phase (between point O and point A). Since the amount of hits which may be obtained before the cutting edges reach the rounding of point A is very low (in proportion to the number represented by the stretch A–B), it is recommended *not* to grind the die plate or the punch until a sharp, zero-radius edge is obtained (level III of Fig. 15-2) but only up to an intermediary level (II) where a small radius still remains. It has been found that when this is done, the number of resharpenings that can be obtained from cutting dies is almost *double* that of the conventional sharpening methods.

For example, the report of an actual case history gives the following results. With a certain die plate it has been found (several times) that if it were resharpened after 75,000 to 80,000 stampings, only 0.005 inch maximum had to be ground away, while if they waited until 100,000 hits, it was necessary to remove 0.010 inch or even more. Consequently, by grinding the same amount of steel (0.010 inch) in two separate operations, instead of only in one, the output (i.e., useful life) was increased by 50–60%.

Sharpening of Punches

It is an established fact that punches are subjected to more wear than die plates. This stands to reason since a punch penetrates the stock and afterward must be pulled out of the material. (In a die plate, normally, the material is in contact with the die opening only once.) In addition, the stock clings to the punch sides like a press fit. Practical observations have demonstrated that punches wear about three times as fast as die plates. Consequently, these practical recommendations should be followed:

1. Do not employ the same kind of tool steel for both cutting members, punch and die plate. It seems logical—and actual practice confirms its correctness—to use a better quality tool steel for punches, with higher wear resistance characteristics, than for die plates.

2. The standard practice to sharpen die plates and punches at the same time, does not make much sense. Make the punches longer than strictly necessary and sharpen them more frequently than the corresponding die plates (two to one or even three to one).

Remove Feather Edges

Among other basic factors that determine the useful life of a cutting die, there is one of practical character that concerns only the press shop people and not the tool designer, nor the toolmaker. The detail in question consists of the fact that after grinding a die plate or a punch, feather edges are formed around the cutting edges. If these small steel slivers are not removed, they are introduced automatically (during the regular cutting with the die) between punch and die plate opening and quickly cause a premature dulling of the cutting edges and sometimes even their nicking or chipping.

There is a very simple remedy for this problem; it is exactly what smart machinists do with the cutting edges of ordinary cutting tools (tool bits for turning, shaping, etc.). After sharpening a die plate or a punch, remove the feather edges with a fine oil stone. It is surprising what a favorable influence this little precaution has on the useful duration of the dies. A fringe benefit of this consists of improved surface finish of the cut surfaces.

Protective Coatings are Helpful

According to corresponding reports, when blanking or punching sheet metal coated with either paper or plastics, both the burr formation and the wear decrease notably on those tool members that are in contact with the coated side of the stock.

This detail is especially significant in the production of laminations for electrical devices. These stampings, made from highly abrasive silicon steel, usually cause very rapid wear of tool members.

Demagnetizing Tool Members

Tool members are usually held in magnetic fixtures for resharpening. The residual magnetism can be a very troublesome problem, and in particularly severe cases it can make the whole operation of the press impractical or, at least, inefficient. In fact, the adherence of metallic chips, flakes, or slivers to the cutting edges of the die produces a rather effective abrasive action causing premature dulling of the cutting edges. The results are shorter runs between grinds and an overall shortening of the tool life. Therefore, carefully demagnetize all sharpened tool members before assembly of the dies.

Special Heat Treatments

Delicate blanking and piercing die plates sometimes develop fine hair cracks after limited use, owing to brittleness provoked by constant strain or repeated stresses. In order to avoid premature wear, it is recommended that these die plates be taken out of the press after a certain number of hits and held at a temperature of 350°F (177°C) for about 1 hour. This heat treatment relieves the stress, without impairing hardness.

Breakage of Die Plates

It happens sometimes that during the operation of a die, the die plate cracks or breaks. This is due either to wrong design (which produces a structurally weak component) or to overload or a combination of the two (which is the most frequent case).

The design of die plates belongs to the province of tool design, which is outside the scope of this book. On the other hand, overloads are caused by the wrong use of the tool and this is a responsibility of the production people.

There are several possibilities of creating situations where a die plate is subjected to overloads that may result in its failure. Here are a few causes:

Starting with the setting of the tool, the necessary parallels needed for supporting the tool must have ample width and should be located as near as possible to the clearance openings of the blanking and punching stations in order to obtain the greatest support for the die and prevent sagging of the die plate (Fig. 15-3). If parallels are set too far apart, the die plate is apt to bend under the force of the cutting action, thus decreasing the punch-to-die clearance. Consequently, there is a definite increase in the required cutting force that in its turn causes further bending of the die plate with further decreasing of the punch-to-die clearance. Such phenomenon often results in severe overloads and die plate breakages.

During the operation of a tool the following details may directly or indirectly cause die-plate failures:

1. Every cutting tool has a fixed punch-to-die clearance that is necessary for a given stock of particular characteristics, chiefly thickness. This value should not be used for a material that requires a different punch-to-die clearance.

2. Inaccurate feeding causes the cutting of half-blanks or half-holes results in the deflection of punches and the chipping of cutting edges. Two strips fed simultaneously greatly overload the die. Similarly, two blanks to be punched in second operation dies may stick together if too heavy a lubricant has been employed for the blanking.

3. In the punching of small holes, slugs often clog the clearance holes, resulting in punch breakage and damage to the die plate (see corresponding item in this chapter).

4. Finally, standard maintenance and inspection practices contribute to a long die life and to avoiding problems. Cutting edges should be checked frequently for sharpness and reground if necessary. In fact, very dull cutting edges mean overloading and thus spalling and breaking of die plates.

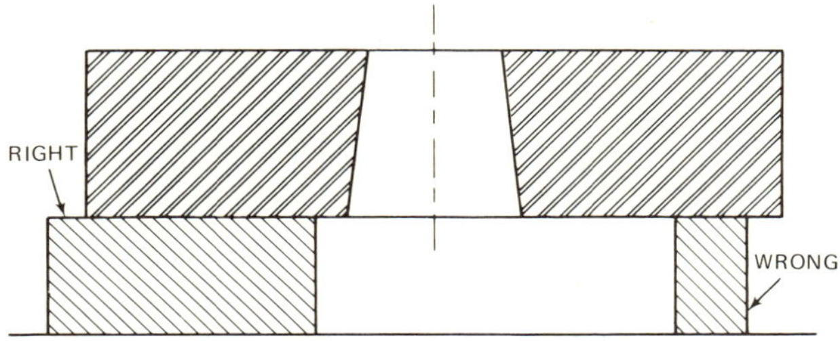

Fig. 15-3. Correct and incorrect die setting (die plates).

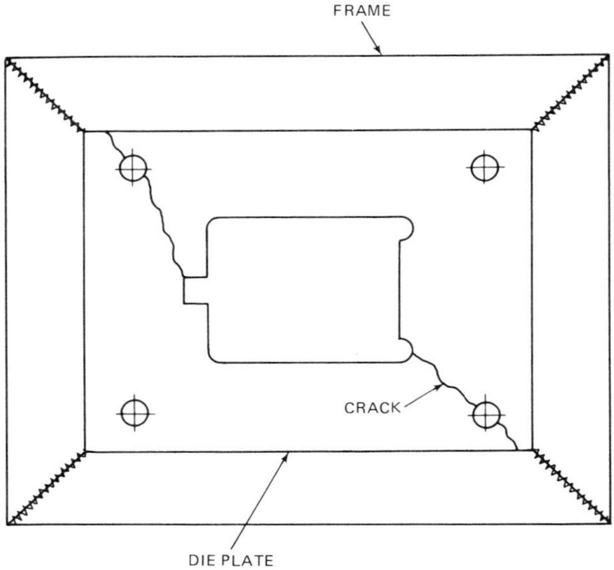

Fig. 15-4. Salvaging cracked die plate.

Repair of Split Die Plates

A cracked die plate does not always mean that it has to be scrapped. Often it may be salvaged, if there is no chipping of the cutting edges. The simple, practical remedy for such cases is a sturdy die-holder frame (made from tough machinery steel) where the two or more parts of the broken die plate are assembled by shrinkage fit (Fig. 15-4) or—leaving enough space for it—by means of shimming blocks. In this way, a proper sectional die plate is produced which, however, functions as a solid die plate.

Burrs

Everyone connected with the production of metal stampings is constantly concerned with burrs. Metal stamping designers, press-tool designers, toolmakers, press operators, inspectors are all burr-conscious. However, curiously enough, many of these people are not able to define clearly the word "burr." Therefore, let us start by giving an explanation of what a burr is and how it comes to be.

Burr is an unwanted, sharp, uneven ridge of protruding roughness around a hole periphery or a blank outer contour, caused by displacement of metal as a consequence of a cutting operation.

It happens quite often that component or part or product designers specify stampings "without any burrs," which demonstrates their lack of knowledge about metal stamping technique. In fact, it is an established fact that in every cutting die as well as in metal shearing machines, the very cutting action causes a more or less progressive dulling of the cutting edges. Consequently, after a certain number of operations, the outer contour of blanks and the periphery of the punched holes throw up a more or less pronounced burr. This is a natural phenomenon and there are no exceptions to this rule: "There is no possibility of producing totally burr-less stampings, at least not consistently."

Burr height is usually measured indirectly (see Fig. 15-5). Burr height b is equal to h (total height) minus t (original stock thickness).

While the burr height remains below certain limits (see below), it does no harm, but when the burr height becomes excessive, the tool must be reconditioned, i.e., the sharpness of the cutting edges must be restored by grinding.

Burr height is practically the same on the periphery of a hole as on the corresponding slug or stamping. For practical reasons of easier handling, burr is usually measured on the slug.

Burr-Height Standards

The quantity of possible hits performed between regrindings by a cutting press tool is determined usually by the admissible maximum burr height on the stampings produced by the tool in question. There is no general consensus about these admissible height limits; it is different in each case, in each company, and even in each stamping. It depends on various circumstances. Here follow some of the main factors: function, utilization, and employment of the part; aesthetics (visibility of the burr in the assembled product); type, magnitude, and quantity of further processing (second operations; assembly and allied operations; etc.) to which the workpieces will be submitted; considerations of safety to the operators (finger and hand injury hazards); considerations of surface finish (scratches of stamping surfaces); considerations of tool member surfaces; storage (stacking); etc.

For example, in the case of laminations for motors and transformers, burrs are especially undesirable both because of the possibility of magnetic "shorts" and because they reduce the space utilization factor, the stacking of the pack.

As mentioned before, there are no industry standards for allowable burr heights. The usual commercial practice is to accept a burr up to the following values:

1. 10% of material thickness for light gage stock;
2. 5% of material thickness for heavy gage stock;
3. in no case more than 0.006 inch;
4. on the other hand, 0.001 inch is always accepted, even if it is more than 10% of stock thickness.

In the majority of cases, burrs of such heights as specified above may be left on the stampings, unless otherwise specified on the production sheet; usually they do not cause any trouble. If they are not acceptable, their removal is easily performed automatically, quickly, and inexpensively by tumbling deburring, vibratory deburring, etc.

Criteria for Maximum Burr Height

Burr removal is a supplementary operation and is moderately expensive, especially in certain cases. However, burr height alone is not a definitive criterion for ease of removal, i.e., for the possibility of removal by tumbling, which is the less expensive and the most popular method of burr removal. For example, it is much easier to remove a high and thin burr (Fig. 15-6) than a low and wide one (Fig. 15-7). Conse-

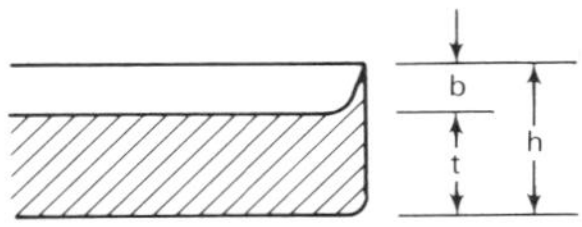

Fig. 15-5. How to measure burr height.

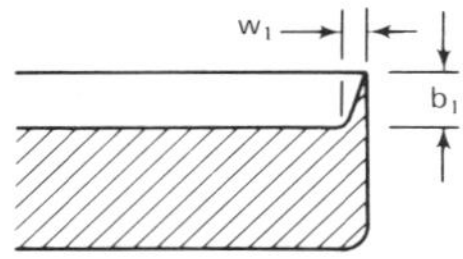

Fig. 15-6. High and thin burr is comparatively easy to remove.

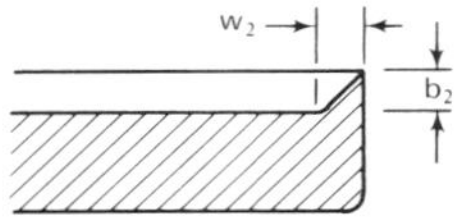

Fig. 15-7. Low but thick burr is more difficult to remove.

quently, the best policy consists of not reaching the point where the burrs must be removed, but controlling their formation so that they are held below certain, reasonable, acceptable values. This burr control is done through preventive means and maintenance, such as frequent resharpening of the tools.

The maximum permissible burr height is sometimes determined by other factors than the function of the stampings. In many cases, the stamping could work efficiently with a comparatively high burr, but, nevertheless, the tool is reconditioned earlier because several production equipment maintenance advantages are obtained, such as: reduced tool wear, shorter grinding time, and less danger of serious tool and machine breakdowns.

Inspection

Burr height must be checked constantly during the production of stampings. Press operators, foremen, inspectors, possibly also diesetters, must check burr height frequently in order to be able to take the necessary steps at the right moment whenever the burrs become too high.

How to Get More From Drawing Dies

The efficiency of a given forming die (longest useful life) is determined by the ratio between its cost—first cost plus maintenance expenses—and the total quantity of acceptable workpieces produced by it.

In a standard drawing die, there are three main tool members: the die ring, the punch, and the blankholder. Apart from mechanical short cirucits (breakdowns), it is the natural wear of the die ring that causes eventual need for replacement. The cause for wear in a drawing die is the continuous friction between the stock and the tool members. However, the punch is subjected to comparatively little or no friction with the stock, only at the stripping; the blankholder may be reconditioned indefinitely, within reason; but the die-ring can be polished only so much, after which its dimensions are enlarged beyond the tolerance limits.

It is the purpose of this section to show the main factors that—either directly or indirectly—contribute to the increase of the useful life of a drawing die ring.

Determining Factors

In order to reach the best tool efficiency in a drawing die, everyone connected with the matter must contribute his share: the part designer, the tool designer, the toolmaker, the press shop foreman, and even the press operator.

The chief determining factors of useful life may be listed as follows: stock characteristics, shell shape and size, tool design and building, stock handling, stock lubrication and die operation.

Since the details concerning product and tool characteristics are outside the scope of this book, only those details that depend on actual production of the drawn shells will be considered.

Press Selection

For any kind of die, but most especially for unguided (plain, open type) ones, i.e., those not mounted in die sets, the selection of a satisfactory press of adequate capacity (at least with a 50% margin, but better with 100%) and maintained in proper working condition, is an extreme importance from the standpoint of increased useful tool life. Almost every press, at the instant of stamping impact, deflects somewhat (especially the very popular C-type presses) misaligning the punch a little. If this deflection is very small, then no harm is done; on the other hand, if the deflection is too pronounced, then the misalignment becomes too great so that uneven wear is produced and thus tool life is greatly reduced.

It must be noted that by reducing the drawing speed (feet per second) the tool life is somewhat increased. In fact, for low speeds, the metal flow is facilitated and thus the strain on the tool members is decreased.

Tool Setting

The skill and care of the die setter play an important role in tool performance. Proper setting of the tool in the press is of utmost importance not only for unguided dies, but also for guided ones (direct guide or die set mounted). In fact, if the alignment (adjustment or centering of the movable and stationary members) is not correct, certain dangerous friction and lateral stresses are set up that provoke premature wear on the tool members.

The correct adjustment of the blank-holding pressure directly affects the tool life. In those cases where the blank-holder must exert too a high pressure (greater than strictly necessary for the operation) premature wear of the working surfaces is the consequence.

A tool mounted in the press, which is not clamped correctly, is in serious danger of becoming misaligned, even if it is correct at the beginning; this is because of shifting due to vibrations. The consequences are premature wear of the working surfaces.

Stock

The stock to be stamped should be clean, without scale, oxidation, or rust. In fact, if hammer or roll scale is present on the metal surfaces, the active working surfaces of the tool members will wear down quickly, because of the strong abrasive action of the hard scales. Oxidized or hot-rolled sheets drastically decrease the total tool life.

In addition, the metal sheet must be perfectly uniform in texture, hardness, thickness, surface finish, etc., because otherwise the unsymmetrical loads caused by such irregularities throw the punch out of alignment and the unavoidable consequences are premature wear of the working surfaces of the tool members.

During the handling of the stock, every care must be taken in order to keep the sheets and the strips perfectly clean and to avoid, at the same time, any bumps or blows which may cause surface imperfections. These imperfections and/or foreign matter deposited on the stock result in accelerated wear of the tool members.

Stock Lubrication

In drawing operations, lubrication creates a film between the stock and the tool members, thus allowing the stock to slip easily between the active tool members. In this way the friction between the working surfaces of the tool and the stock that is being processed is reduced and, therefore, the heat produced by the operation is greatly decreased. Consequently, less wear on the tool components and thus longer life is achieved.

The type of lubricant has a great influence on results. Always consult the supplier and make exhaustive tests in order to find out which lubricant gives the best results for each operation. For every type of operation and every kind of stock there is a "best" lubricant; there is no such thing as a "universal" lubricant, i.e., there is no single type of lubricant that is equally satisfactory for all kinds of press operations and all kinds of stock.

Pressroom

The pressroom may also contribute considerably to the tool performance. One of the pertinent details consists of the periodical cleaning of the tool in order to avoid accumulation of dirt, chips, and foreign matter, which, when mixed with the lubricant, act as an abrasive and considerably increase tool wear.

Finally, the press operator may contribute directly to the prolongation of the useful life of the tool by

a. avoiding any irregularities (double blanks put into the nest, faulty nesting of the blanks, etc.);

b. informing the foreman immediately as soon as wrinkles or cracks appear in some shells;

c. in general, giving the tool the care it deserves as a precision piece of equipment.

Problems with Slugs

Automatic and semiautomatic production of metal stampings involve much more than simply providing expensive presses, dies, and material-handling equipment. It also means assurance of trouble-free operation. As mentioned in this book repeatedly, unscheduled production interruptions must be prevented at any cost.

One of the ever present problem areas in the production of sheet metal stampings is small slugs.

Punching and blanking are essentially the same operation. In both cases a continuous object, the slug, is cut out from sheet metal. The shape and dimensions of the slug are determined by the die opening. The basic difference between punching and blanking consists of the use of the slug. In punching the slug is called a *punching* and is scrap; in blanking it is called a *blank* and is the actual workpiece. However, there are occasional cases where the punching from a blank becomes an independent component (typical example: rotor laminations are punchings from stator laminations).

Slugs sometimes cause problems which will involve expense, loss of time, labor costs, and lowering of the production efficiency of the corresponding press tools. There are two major inconveniences caused by slugs that must be taken care of: clogging of die openings and slug lifting or pulling.

Jamming the Slug Discharge Holes

Jamming or plugging of the slug discharge holes in the die plates and/or in the die shoes is due to the successively cut slugs becoming compressed together and building up into almost solid columns. Therefore, under certain unfavorable conditions, some slugs in the column can shift to a crosswise position and literally clog the slug exit holes in die plates and/or die shoes. The possible consequences are quite numerous and varied: punch bending, punch breakage, die plate cracking, damage to the die shoes, and even damage to the press.

Slug Pulling

Slug pulling is a return of slugs above the die plate. Comparatively small and medium size, regularly shaped, thin, flat slugs have a decided tendency to cling to the bottom end of the punch and to be lifted from the die opening at the return stroke of the press ram. There are several disagreeable consequences: possibility of misfeeds, subsequent blanks are more or less marked and deformed; sometimes if the slugs shift onto some cutting edge, punches and/or die opening edges may be damaged, and punches may break.

Causes and Remedies

Experts on press working of sheet metals (press-tool designers, toolmakers, press-shop foremen, etc.) have various and sometimes opposite opinions about the possible and probable causes and corresponding remedies of these two problem areas. Therefore, the general consensus of opinion about these conditions is presented here.

First of all, it has been found that there are several factors that may cause either clogging of die openings or the tendency for slug pulling. Generally, there is no single cause in a given case, but a combination of causes occurring simultaneously.

The chief causes of trouble with slugs are these: incorrect die design, inadequate die making, incorrect punch-to-die clearance, advanced tool wear, excessive and/or inadequate lubrication, and unfavorable stock. The corresponding *remedies* for preventing or eliminating the problems belong chiefly to the province of the product and tool engineer, so they are outside of the scope of this book. Here only a few details are dealt with which are the direct responsibility of the production people.

Tool Setting

Sometimes it is the die setter who is responsible for the clogging of the die openings. It is obvious that the tool must be so located on the bolster plate and/or the press table that the slugs have an absolutely free discharge path, i.e., all the clearance holes of the die plates and die shoes, must have correctly positioned discharge holes of the right size. Quite often riser blocks and/or parallels partially obstruct some discharge hole; the consequences are almost always a jamming of the hole in question.

In addition, the shut height of the press should be so adjusted that the slugs are pushed deeply into the die cavity, in order to avoid slug pulling.

Tool Wear

The keenness of the cutting edges—both on the die plate and on the punch—is of great importance. In addition to other considerations of tool life, appearance, and function, burred slugs have a greater tendency to cling to each other and form columns, which then obstruct the die openings. Consequently, for best results, slugs should always have minimum burrs.

Also, if the punch bottom end is deformed by wear, becoming slightly mushroomed, burring increases, and the slugs increasingly tend to stick to the punch and lift from the die.

Lubrication

Stock lubrication must be studied and controlled carefully. Experts on press working of metals strongly recommend that the right kind of lubricant always be used.

Avoid the use of lubricants that are too viscous, which increases the tendency of slugs to jam the die openings and to pull. Therefore, when one of these problems occurs, check among the other details the lubricant viscosity and, if possible, change to a less viscous lubricant. Also remember that lubricants must always be applied sparingly. Too heavy a lubricant film acts as a kind of cement and increases slug problems.

Magnetized Punches

Punches sometimes become magnetized, either while being sharpened or simply by the earth's magnetic field combined with the punching action. Of course, this phenomenon increases the danger of slug lifting. Therefore, always demagnetize punches after sharpening and whenever there is a suspicion that they have become magnetized naturally.

Unsticking Jammed Presses

Occasionally, an operator makes a double hit—two blanks are put into a die simultaneously—causing the press ram to stop on dead bottom of the stroke. Sometimes it is not the operator's fault: The pitman screw locknut becomes loose so that the pitman unscrews very slowly, thus shortening the shut height of the press. Other causes for press jamming may be exceptionally heavy stock thickness, sudden interruption of electric power, or too deeply set shut height during die setting.

There are several proven, practical solutions for this problem. These solutions are divided into two separate groups: remedies and peventive methods—or in other words, what to do after or before the problem arises.

Remedies

The most usual situation is a punch press jamming unexpectedly, i.e., without any preventive steps having been taken. Here are the remedies to try, presented in a roughly increasing sequence of difficulty: (Note, of course, that in every case the die and punch clamping straps and bolts must be removed before there is any attempt to unstick the jam.)

1. If the press is equipped with tie rods, loosen them. Quite often, this is sufficient. If the nuts are too hard to turn, heat the tie rods in order to expand them.
2. Gradually apply to the die members external forces, such as wedges, hydraulic jacks, etc. Never use impacts!
3. Take off the gibs, which guide the press ram or slide, and try to slide the ram together with the die out of its seat, in a forward direction. For this operation a long rod, a crowbar, or something similar is used; never a hammer or a mallet.
4. Put dry ice around the tool and around the pitman screw to make them contract, thus reducing the jamming pressure.
5. For steel-framed presses, heat the frame in order to expand it. Do not do this with a cast frame, because it might crack.
6. Use any reasonable combination of methods 1–5.
7. If none of the above methods work, then it must be decided from an economic standpoint what must be sacrificed, the tool or some press member. Usually it is preferred to cut through the pitman screw, because the screw, having taken a tremendous impact, is deformed anyway or, at least, the threads are damaged and it will be necessary to replace it. In addition, the cost of a new pitman screw is almost always much less than that of a die.

Preventive Methods

In many cases, if a shop has had comparatively many press jammings, the responsible people should try to take some steps *before* a critical tool is set up in a press, so that the unsticking of the press can be easy, quick, inexpensive, and safe for the tool, the press, and the workmen. Here follow several ideas, practically all of them tried out successfully in actual practice.

1. Put a sheet of heavy paper or cardboard between the die bottom and the bolster plate and/or between the punch-holder top and the ram bottom face. Then, in case of jamming, burn out the paper.
2. Instead of paper, use a sheet of rubber. In many cases the yielding of a rubber sheet (⅛ to 3/16 inch thick) compensates automatically for overthicknesses or double thicknesses of stock and thus any jamming is avoided.
3. Machine a cylindrical recess in the ram top to hold a strong flanged cup (Fig.

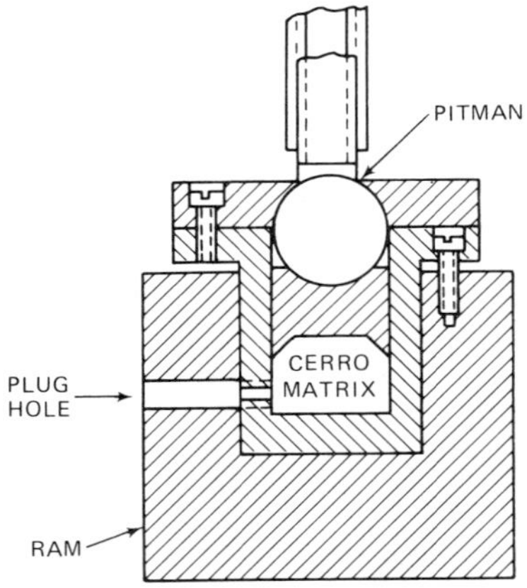

Fig. 15-8. Low-melting alloy used for unsticking jammed presses.

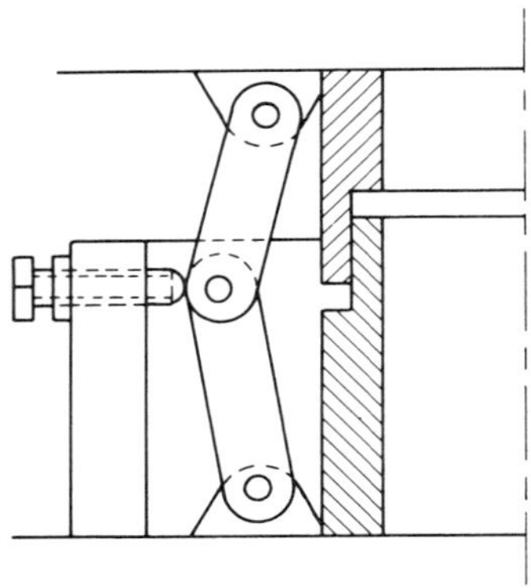

Fig. 15-9. Toggle links for unsticking jammed presses.

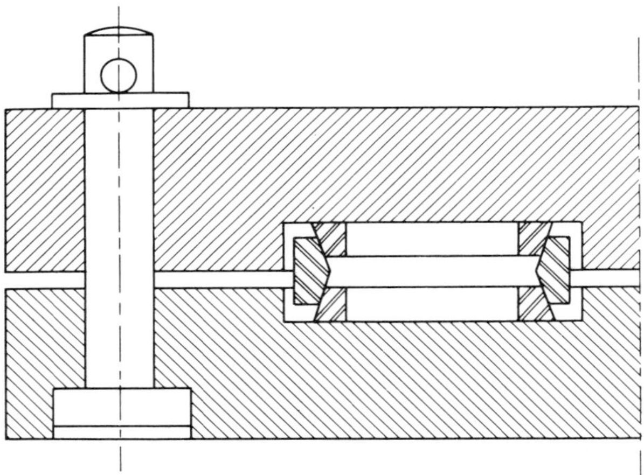

Fig. 15-10. Overload-compensating composite bolster plate.

15-8). In the interior of the cup place a cake made from Cerromatrix or from an alloy of tin and lead (50–50). Rest on the cake a properly shaped steel cylinder that will receive the thrust from the ball joint and the pitman screw. In case of jamming, take out the two or more plugs of the clearance holes, heat the cup so that the low-melting alloy softens or melts outright. Thus the jamming pressure is removed.

4. Toggle links. As shown in Fig. 15-9, the punch holder is of composite, telescopic design. The two parts are connected with three or more toggle links held in place by means of a ring having bolts that keep the links in almost a straight line. In case of jamming just loosen the bolts.

5. There are several designs of "safety fuses" for presses and press tools. These protective devices are either expendable components that break in case of overload (shear plates, shear pins, etc.) or spring-actuated mechanisms or hydraulic or pneu-

matic mechanisms that yield under the influence of overloads and thus avoid jamming and breakage.

6. Among these overload protecting devices a comparatively simple springloaded kind of bolster deserves special mention. As shown in Fig. 15-10, it is composed of two substantial plates, separated by three or four ring springs (each of them composed of two half inner rings and one outer ring, or vice versa), symmetrically positioned. The springs are held in preloaded condition by four or six substantial bolts and dowel pins or any other similar arrangement. As long as the load on the bolster is within the preestablished limits, it behaves like a solid block. However, when the pressure becomes higher, the device collapses so that no jamming occurs.

All the solutions presented above are "homemade," i.e., any shop or toolroom can make them. In addition, there are several commercially available products for the solution of the problem in question.

Press Selection

A very efficient method for avoiding jamming of presses consists of properly selecting the right kind and size of press. For example, for heavy stamping and coining operations use screw rather than crankshaft presses, because they cannot be stopped on dead bottom and can always be reversed easily.

Magnetized Stampings

The magnetization of steel stampings can have harmful consequences for both the product and the press tools. Product damage occurs when tiny particles of steel cling to magnetized stampings and score mating surfaces or, in severe cases, jam moving parts. Even without picking up steel particles, magnetized mating parts tend to stick together, restricting the intended free movement.

Furthermore, the efficiency of barrel finishing or plating is impaired if parts are held together by magnetism. And, when magnetized parts are utilized in electromechanical or electronical devices, a constant, high-speed movement may induce electrical current in adjacent conducting material.

It should be kept in mind that residual magnetism is created only in paramagnetic materials such as steel, nickel, or cobalt. The lower the content of carbon and other alloying elements, the less the intensity and duration of residual magnetism—magnetism in low-carbon steel stampings decreases rather quickly if no further magnetizing action takes place.

Residual magnetism in press tools can be very troublesome. In severe cases, it can make operation of the press impractical or inefficient at best. Although the stock is in contact with magnetized tools for only a short time, it may become magnetized and cause the difficulties mentioned above. Furthermore, the minute metallic chips, slivers, flakes, etc., that cling to the active surfaces of forming dies can scratch the surfaces of highly finished stampings. The same material causes premature dulling of punching and notching dies. Shorter runs between regrindings and an overall shortening of tool life result.

Finally, strongly magnetized punches have a tendency to lift small slugs and blanks out of the die plate opening and deposit them on the die plate top surface.

This can cause problems ranging from stamping damage to punch and die plate breakage.

Origins of Magnetism

Stampings may be magnetized in several ways. Steel sheet, coil, and strip handled with magnetic lifters may retain residual magnetism, or a prior operation such as lamination work on silicon steels may impart the same condition.

In the press, magnetizing can occur from electrical charges induced by the blanking or punching operation, even though the tools themselves may not be magnetized.

The sources of tool magnetization are numerous. The main cause is the use of magnetic chucks during finish grinding and regrinding operations on die members. This can occur even though the current has been reversed in the magnetic chuck, because tool steels have higher magnetism-retentive power than low-carbon steels.

The same magnetization can occur without the use of magnetic chucks. The earth's magnetic field, aided by the vibration of the tool as it works, can create the same situation. In this respect, the physical location of the press has a certain influence. If a long machine such as a press brake is lined up with the earth's magnetic field, and if it is used in a rapid, repetitive operation, the tools mounted in the press are more likely to become magnetized than if the press were aligned east to west.

Occasionally, magnetism is caused by static electric charges. These are usually picked up from the V-belts used to drive the flywheel and are transmitted from the press to the dies by direct contact. A final cause is exposure to strong magnetic fields. Location near any apparatus which produces strong electric or magnetic fields, such as high-current-carrying conductors, will tend to increase the chance of the tool picking up magnetism.

As with stampings, the phenomenon of residual magnetism occurs only if the die members are made of paramagnetic metals. If they are made of carbides, Kirksite, or other special nonferrous alloys, there is no possibility of magnetizing the cutting members of the tools.

Prevention

Residual magnetism may be combated by preventive or corrective measures. Prevention is the most desirable method, but it is also the most difficult, because there are only a few approaches available:

Stampings. Avoid magnetic lifters for handling stock, use hook lifters instead; if possible, change the type of steel to one with less carbon content and/or alloying elements.

Tools. If possible, employ nonmagnetic materials for die members; reorient the press to avoid the influence of the earth's magnetic field.

Presses. For press construction, use a mild steel that is magnetically "soft" and has a low hysteresis—any magnetism which may be introduced in manufacture or during use will dissipate fairly rapidly.

Remedies

Corrective means for magnetized parts are selected according to the circumstances, objectives, and characteristics of the components. The most popular technique for stampings, tools, and press components is to pass them through a demagnetizing device, which is essentially an electric coil energized with ac current. This creates an alternately reversing magnetic field that completely removes the residual magnetism in the parts when they are placed in the opening of the coil. Of course, the magnetic field created by the demagnetizing coil must be stronger than that of the magnetized part. One of two procedures may be followed:

1. The power is turned on full before the part is introduced into the coil opening. The part is passed through slowly, sometimes after a short dwell in the interior of the coil, and is removed at the other side of the coil. The power is turned off after removal of the part. (A sudden turning off of the power while the part is still under the influence of the magnetic field of the demagnetizer may increase, instead of eliminate, the original residual magnetism.)

2. With the power turned off, the part is placed into the demagnetizing device. The power is turned on full. After a period of time, the power is decreased gradually until the field is decreased to zero.

For efficient processing, small stampings are demagnetized in groups with the aid of trays or tote pans.

Demagnetizing equipment is commercially available. However, when circumstances do not warrant the purchase of such a device, one can be easily and economically built.

Sometimes it is possible to resort to the simple expedient of annealing the parts at 1400°F (760°C) and allowing them to cool slowly. This technique can be used only in case of stampings, raw material, and some press parts, but never die members, and then only in cases where decrease in strength and rigidity is not objectionable. In some instances, the stress-relieving benefits of annealing make this a preferred method.

When annealing rather large magnetized parts, it is good practice to let them cool in such a position that their major axis is in an east-west direction.

V-belt-induced static electricity can best be avoided by installing a soft copper wire brush to remove the charge from the belts by running the wire to ground.

Detection Methods

The presence of residual magnetism is usually quite evident visually. However, it is good practice to apply some preventive measure before magnetism becomes objectionable. Therefore, it is necessary to check the stampings, the tools, and the press periodically for the presence of this condition.

Detection may be accomplished with a special instrument built for the purpose, but for many shops, a small magnetic compass is satisfactory. By the deflection of the needle, the presence of magnetism is immediately indicated. Parts must be longer than the compass needle for this method to be effective.

Distress Signal

In a press shop the noise level is quite high. Consequently, verbal communication is rather difficult and requires a special effort. Therefore, in some shops special signaling methods have been devised for achieving efficient, reliable, basic understanding between the working force and the foreman, especially when problems occur.

The simplest solution consists of a large disk (about 10 inches in diameter) made

Index

A

B

C

D

E

I

J

M

N

P